Urban Trees

A Practical Management Guide

Urban Trees

A Practical Management Guide

Steve Cox

THE CROWOOD PRESS

First published in 2011 by
The Crowood Press Ltd
Ramsbury, Marlborough
Wiltshire SN8 2HR

www.crowood.com

British Library Cataloguing-in-Publication Data
A catalogue record for this book is available from
the British Library.

ISBN 978 1 84797 298 9

Frontispiece: This veteran beech in Montpellier
Gardens, Cheltenham, is much older than
the gardens, which were laid out in the early
nineteenth century.

Typeset by Bookcraft Ltd, Stroud, Gloucestershire

Printed and bound in India by Replika Press Pvt Ltd

Dedication
This book is dedicated to Jane, Molly and
Rowena, without whom it would have been written
years ago!

Acknowledgements
The idea to write a book was seeded into my
mind during a convivial chill-out evening with
Graham Gover on 1 January 2000. (It has taken
ten years, but I told you I'd do it, Graham!) It
is a personal view of arboriculture in the UK.
However, I have received much support from
fellow arboriculturalists. In particular thanks go to
Jonathan Fulcher, Jerry Kelsey, Jo Ryan, Jeremy
Barrell, Jim Hillier, Neville Fay, Nick Eden, Brian
Simpson, Keith Sacre and Glynn Percival. Many
more people have contributed to the book
through conversations over the last two years.
Special thanks go to Richard Stamp for checking
my text and providing ongoing good advice and
encouragement.

Contents

Introduction

Trees are a vital part of urban life. Trees grow in towns and cities and are either planted, grow up uninvited, or become engulfed by the expansion of man's work and home environment. This has been true since people first began to congregate into settlements and so the journey together of man and tree has a long history.

City living, although it has been around for millennia, has only become the dominant experience of people around the world in the last few decades. The intensity of this lifestyle has put immense pressure on those trees that serve us within it. In the past, it may have been possible for city dwellers to pay relatively little attention to trees, because they were always arising in some corner of the plot, or they were abundant not far away in the countryside. However, the twenty-first-century UK experience in town and city is that the trees we have are constrained and restricted by all manner of things. The development of management regimes and maintenance programmes is therefore vital to ensure that these trees remain positive elements in the cityscape and also that there is effective consideration and effort given to establishing the next generation of trees.

Trees are connected to the past and the future as well as the present. We manage, protect and are responsible for those trees that arose in previous decades, maybe centuries; we respond to the present problems arising from the close proximity of people and trees and we also make decisions that will affect the way that trees contribute to city life in the future.

In the UK, the population of 62 million is almost 90 per cent urban. The growth of towns and cities has progressed organically, in a haphazard way that urban planners and designers have struggled to harness, and it is easy to find places where opportunities for tree planting have been either missed or mishandled.

Town expansion was at its most intense in Victorian times, when land was gobbled up for new cities that were needed to house the swelling workforce coming in from agricultural areas. Obviously, the most thought and effort was spent on the richer parts of the boroughs, but a valuable legacy of large existing trees and new trees planted was passed down into the twentieth century. Land managers were often deeply experienced men who knew how to plant, grow and manage trees. This expertise has been severely battered during the past century as successive national events have changed the way our natural resources are managed. World War I led to many woods near to towns being chopped down, while World War II resulted in widespread damage to many cities. Lack of funds and experienced manpower meant that planting and pruning programmes had to be curtailed or abandoned and reorganizations led to a disconnection from known and proven ways of managing public open spaces and resources.

However, within this broad-brush picture many positive features have come out of our history over the previous century. The growth of the garden city concept and its spread into new towns has meant that open space management and tree planting have been seen as important elements in any development of significant sized areas. Engineers have

Trees in urban areas will inevitably encounter humans.

always thrived on problems that need to be overcome and there are many ways in which the interconnectedness of people and trees in towns and cities needs an open and thoughtful approach so that infrastructure and trees are not damaged. The planting of trees in streets or on roofs are examples where innovation can help to improve the quality of our urban landscape.

This book sets out the basics of arboriculture, with the emphasis on urban areas. Tree management has always been needed wherever people and trees have lived close together, but now that need is greater than ever because of the complexity of urban life. Not only do we need to understand the ways that trees help to make cities bearable for people, but we also need to understand what trees need to make cities bearable for them. Only by succeeding in providing an environment that allows trees to thrive will mankind be able to get the most from the presence of trees in towns.

In the UK, this challenge is intense. As well as having a world-renowned resource of old, mature trees, we also have a dense network of built infrastructure that leaves almost nowhere untouched. If we do not develop further a culture of looking after trees in towns over the next couple of decades, we will be left with very little from the legacy passed down to us from the Victorian and Edwardian town-builders.

The book begins by looking at what trees are and what they do for us, then switches to towns and briefly traces the development of urban areas in Britain. After considering how these urban areas now affect and mould trees in towns, we look at the issue of planning urban tree management. Chapters on planting, establishment and maintenance, monitoring and assessment, and two on tree management on different urban sites are followed by a quick outline of current UK law in relation to trees. The last chapter takes a brief look at climate change, the future of arboriculture and trees in urban areas in the UK.

The subjects and details here are not exhaustive. I have not said everything that is to be said relating to urban tree management. In fact, quite a lot has had to be left out. It may be that, with growing interest in this area, any holes can be plugged by subsequent editions or by other books. There is a need for books and information that draw together all of the innovative work and practical wisdom available about urban trees.

REFERENCES

Denham C. and White I., *Difference in Urban and Rural Britain* (Office for National Statistics, 1998).

chapter one

Trees: How They Work and What They Do for Us

WHAT TREES ARE

Everyone knows what a tree is. We all have a mental picture of a 'typical' tree, but, as with many things, once you try to come up with a neat, comprehensive definition, that mental picture turns a bit fuzzy. Most general books on trees glide straight past the definition and instead concentrate on how awesome they can be. They emphasize the size, varying leaf colours and textures and the importance of trees as a home for wildlife. But what exactly is a tree? Where are the edges of 'treeness'?

Fungi share some characteristics with trees; they have extensive root systems and can last for decades in the soil. Vegetables may have a stem structure like a tree (for example, broccoli). Shrubs are very like trees, but are often smaller and shorter lived. Some climbing plants reach great heights (by using trees for support) and they can have thick stems themselves. Grasses may not be anything like trees if we think of a garden lawn, but palm trees are in fact a specialized form of grass and not trees at all. What about a banana tree? That is really a giant herb.

From a botanical perspective, a tree may be any plant with a self-supporting, perennial woody stem (that is, living for more than one year) (see Thomas, 2000). Or it may be a perennial woody plant, generally with a single stem (trunk) (see Raven et al., 1986).

Trees differ from most other plants and fungi, in that they have woody stems. This refers to the secondary thickening of the stem and branches that is produced by the growing cambium just below the bark. Many shrubs and climbers also have woody stems, but they don't have a single, self-supporting stem, or the ability to reach greater heights. The general distinction between tree and shrub that horticulturalists use is that a shrub cannot get much taller than 6m (20ft) and is usually multi-stemmed from close to ground level. Climbers can reach greater heights, but cannot get there without something to hold on to.

Having clarified the definition of a tree, we need to acknowledge that there are still some loose ends. In mountainous regions and close to the poles, there are plants that may be only a few centimetres tall but are considered to be trees and not shrubs. Here, it is the environmental conditions that limit size, so the plants are still regarded as trees in the same way that bonsai trees are trees and not shrubs.

What is the legal definition of a tree? Does the justice system use the botanical definition of a tree, the horticultural definition or something else? In legal terms in the UK, judges have avoided a precise definition of a tree and

A maidenhair tree – a deciduous conifer with broad leaves and squashy fruit! (Photo courtesy of Jo Ryan)

have opted for an 'everyman's' definition. The general definition of a tree used in the courts is that anything generally understood to be a tree is a tree.

What's in a Name?

The terms used to describe trees and wood types can be confusing and ambiguous: in general terms, trees are either broad-leaves or conifers; they can be deciduous or evergreen and are often referred to as either softwoods or hardwoods. Broad-leaved trees have foliage in the familiar leaf shapes. They are broad when compared with conifer needles, but can be all shapes and sizes depending on the species of tree. Conifer trees are characterized by very thin, needle-like leaves that last for more than one year. Conifers all produce some kind of cone fruit, the name conifer meaning cone bearer. Pines and cypresses are typical conifers. In England, the majority of trees are broad-leaves, while most forestry plantations are conifers, although in some parts of the country there are large numbers of pines that change this general picture.

Deciduous trees lose all their leaves in autumn when they stop growing, then sprout new ones each spring. The best-known British trees are deciduous ones such as oak, beech and sycamore. Evergreen trees retain their leaves through the winter, so are valuable for providing screening or shelter. Often leaves carried through the winter are shed in the spring as new leaves sprout. Evergreen oak and eucalyptus are good examples of such trees. Conifers can be described as 'evergreen' as, in general, they do not lose all their foliage at the same time.

However, rules are there to be broken and some conifers are deciduous, while some broad-leaves have cones! But these exceptions do not overturn the general rule. There are very few deciduous conifers. Swamp cypresses and dawn redwoods, larches and the golden larch all have bare branches through the winter. Alder is a broad-leaf tree that bears brown cones each winter which superficially look like conifer cones.

The terms softwood and hardwood are easy to define, but not necessarily intuitively understood. They are carpentry terms rather than botanical ones, relating to the characteristics of the timber, not the living trees. Softwood timbers come from conifer trees and are easy to work with. Hardwoods come from broad-leaved trees and have more chemical ingredients that blunt carpentry tools. But this brings the apparent anomaly of balsa wood (usual density $150kg/m^3$) being a hardwood and yew (usual density $670kg/m^3$) being a softwood, so hardwood should not necessarily be equated with heavy and soft with light. If you have ever experienced a branch fall from a tree, you will know that there is no such thing as a soft wood unless it's well rotten before it falls.

HOW TREES WORK

Growth

The body of a tree is composed of a trunk that connects an underground root system to a branching crown, on which are held the leaves. Growth occurs just beneath the bark of the tree in a thin layer (cambium) that produces a layer of cells on each side (outward and inward), resulting in the thickness of stems and branches increasing. The outer edge of the bark sloughs off as it ages and trunk diameter increases. At the ends of the branches are buds that are programmed for growth and packed tight at the end of summer, so that, in the following spring, they can 'spring' into action and shoot outward. These shoots extend the branches and produce fresh leaves. Sometimes, shoots grow from the trunk or large branches where enough light reaches them to stimulate growth.

The roots grow to form an extensive network of woody structures that anchor the tree to the ground. They pass nutrients and water from their tips to the trunk and from there to the branches and leaves. Roots radiate out from the stem, but are usually limited to the top metre or two of soil. Most roots are found in the top 600mm (24in) of soil.

Growth is powered by sunlight through the process of photosynthesis within leaves. Leaves are highly organized structures with a protective surface above special palisade cells containing chloroplasts, which is where the chemical reactions take place. Within the chloroplasts, chlorophyll, which gives the leaf its green colour, is used to focus the power of the sun's rays onto carbon dioxide and water to produce packages of energy as sugar molecules, which are then combined together to form starch. This is transported through the phloem (veins) around the plant body to wherever growth is occurring and energy is needed – cambium in the trunk and branches, buds and branch tips and in roots and root tips. Oxygen and water are also produced in this process. These molecules find their way between the leaf tissue below the palisade cells where there are spaces connecting to special portals, stomata, on the underside of the leaf. As the stomata open the water vapour and other gases are released into the air. The veins in the leaf operate a one-way system with the phloem transporting the starch away from the leaf while the xylem part brings fresh water and nutrients to the photosynthesis sites.

So, wood is made from this mix of fresh air, water and a sprinkling of nutrients swirled

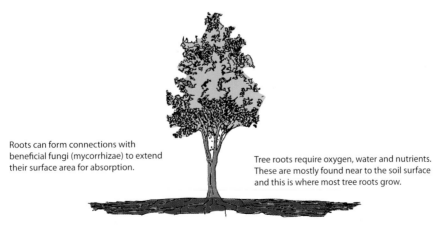

Roots can form connections with beneficial fungi (mycorrhizae) to extend their surface area for absorption.

Tree roots require oxygen, water and nutrients. These are mostly found near to the soil surface and this is where most tree roots grow.

Roots anchor the tree and absorb water and nutrients from the soil.

Root depth is limited by lack of sufficient oxygen, soil compaction and waterlogging.

Generalized tree root system.

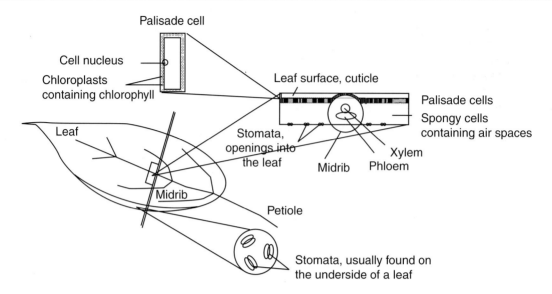

Leaf detail.

tree reorganizes these tiny particles changes them into a hard-wearing, solid material that is amazingly useful to mankind.

There is a distinction between growth and life. Tree cells need to use the energy produced by photosynthesis to keep themselves alive first and secondly for growth. Respiration is the process of unlocking that energy, delivered as starch, either to keep cells alive or to power growth. Staying alive requires respiration and

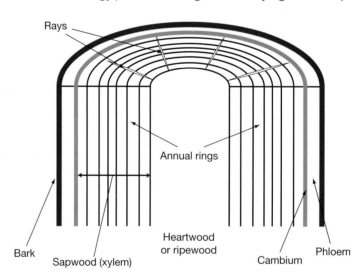

Cross section of a tree trunk. Bark: protects inner tissues from damage and disease. Phloem: transports dissolved starch and hormones throughout the tree. Cambium: a layer of active cells creating xylem cells on its inner side and phloem cells on the outer side. Sapwood: xylem cells that transport water and nutrients from roots to leaves. Heartwood: inner zone of old wood that is no longer used for water transport but adds strength to the whole tree structure. Annual rings: formed by the pattern of growth of cells. Medullary rays: transport nutrients horizontally across trunk and store starch.

all tree cells do this unless they are more useful to the tree dead, such as heartwood in the centre of a trunk. For instance, during the day leaves use light to power photosynthesis and this produces energy, water and oxygen. The oxygen and water are released back into the atmosphere. At night, there is no light to power this process, but the leaves continue to stay alive and to grow. Through respiration, oxygen is used up and carbon dioxide produced and released into the atmosphere. In the daytime, the level of carbon dioxide produced in respiration is less than the level of oxygen produced in photosynthesis.

Reproduction

Trees produce flowers, usually on the branches, that contain male and female parts. The male

Insect-pollinated fruit tree blossom.

A wind-pollinated hazel flower.

parts contain pollen, which is either spread between trees by the action of the wind, or it gets stuck to the bodies of insects as they visit the flowers to feed on the sugary nectar. As the insects visit flowers on different trees, they deposit pollen onto the female parts of the flowers.

The pollen grows down into the flower ovary and fuses with an egg, with a fruit forming around the resulting seed. When ripe, the fruit falls from the tree and is scattered by the action of the wind, or is collected by animals that eat either the fruit or the seed (or both). This results in seeds being moved away from the base of the mother tree. Any undamaged seeds usually lie dormant through the first winter. During the spring of the following year (or possibly one year later), if the surrounding conditions are favourable the seed sprouts and the first root, stem and leaves appear.

Examples of wind-pollinated trees include oak, birch, elm and maple. These trees in general have inconspicuous flowers. Insect-pollinated trees have more showy flowers and include magnolias, cherries, apples and lime. Trees that produce small, light seeds use the wind to distribute them away from the mother tree. Examples include birch, elm, willow and poplar. Alder trees often grow besides watercourses, producing light seeds with an inbuilt air bag that keeps them afloat while they are carried downstream. Larger seeds either fall to the ground and are then collected and 'squirrelled' away in a forgotten hoard (for example, hazel or oak), or the fruit is eaten and the undigested seed is released from the animal (very often a bird) in its droppings. In this way, the seed starts off on the ground with its own portion of fertilizer. The seeds are designed to be spread away from the mother tree. Many trees will not grow beneath the canopy of the mother tree; they need to get away to where there is adequate light from above.

Tree seeds often exhibit some form of dormancy that stops them germinating in the wrong conditions. The dormancy locks the fragile living seed inside a shell, like a time lock on a safe. This time lock is controlled by a mixture of factors, usually temperature, moisture content and daylight, or possibly by a chemical coating that has to be broken down by the outside environment. However, not all seeds are like this. Some are opportunists designed to grow as soon as they come to rest on a suitable surface. Generally, when a tree invests significant resources in protecting its seed, it means it produces fewer of them. Opportunistic seeds that grow anywhere are usually produced in abundance by the mother tree.

WHAT CONDITIONS DO TREES NEED FOR GROWTH?

Air

The air surrounding us and the above-ground parts of trees at sea level is, roughly, 78 per cent nitrogen, 21 per cent oxygen and 1 per cent other gases. Trees need oxygen for respiration and carbon dioxide for photosynthesis. Carbon dioxide is present in the atmosphere at low levels (around 0.04 per cent). Wind movements are usually sufficient to provide enough oxygen and carbon dioxide for plant processes.

Below ground, around 50 per cent of a good quality, undisturbed soil is composed of spaces containing air that is needed for root respiration. The percentage of oxygen is usually lower than above ground and the route by which air is replaced is more convoluted, with it needing to pass from successive soil pores until it reaches the surface. This means that the amount of oxygen decreases with depth and carbon dioxide released by roots in respiration (and respiration of other soil organisms) may stay in the soil pores, further affecting oxygen levels. Soil oxygen levels of 10–12 per cent seem to be needed to maintain satisfactory growth and many trees' roots stop growing when the oxygen level is below 5 per cent and die when it is less than 3 per cent, although there is some species variation. Soil carbon dioxide levels are higher than above ground, being approximately 0.25 per cent, and can rise much higher where there is no effective gaseous exchange with the surface.

Light

Light affects tree growth in many and profound ways, the most obvious of which is its use as the power source in photosynthesis in the leaves. Light falling on a leaf surface produces a number of effects: it is reflected back into the atmosphere either as long-wave radiation or as convected heat, the light warms up the leaf and light energy is used to convert water to water vapour, which then passes into the air through transpiration. Only a small part of the radiation is allowed to pass into the leaf, where special structures, chloroplasts containing chlorophyll, use the energy to break open the molecules of water and carbon dioxide and recombine them into molecules of sugar.

Because of the strength of light, leaves also need protection from it. Leaves are most efficient at using moderate levels of light. Maximum photosynthesis occurs at between one-third and two-thirds of full sunlight. High light levels damage the leaf tissue and low levels provide insufficient energy to power photosynthesis.

Sunleaves form at the top of trees where there is the greatest exposure to light. Their leaf surface has additional waxy or rubbery screening to protect the chloroplasts and other cell contents. In contrast, leaves produced within the crown are exposed to much lower light levels and tend to be larger, thinner and more papery.

Some trees cope with shade better than others and weak specimens within tree groups of all species can lose out in competition with their neighbours, finding themselves declining in growth rate and health and ultimately dying due to lack of light. Even if this decline is accompanied by pest or disease attack, or drought or other damage, it is often the lack of light that begins the process.

Light is also used by trees to control their response to environmental conditions. Onset of winter is signalled to trees by reduction in day length, with trees responding to this by physiologically preparing for cold temperatures and dormancy. But light is used as an environmental cue for many processes within the tree: growth initiation, cambial activity, shoot growth, flowering, abscission of leaves in autumn and seed dormancy factors all have some light-sensitive mechanism at work.

Temperature

Climate zones are characterized by temperature. Temperatures drop with increasing latitude, toward the poles and with altitude. The important temperatures for tree growth are the upper and lower limits to growth and the optimal temperature. The rates of physiological processes are closely linked to the action of enzymes and generally increase from near 0°C to a maximum in the range of 20–35°C, dependent upon species.

Trees are considered to be well adapted to their location when the climate allows them to carry out all their physiological functions for growth. First, a tree must be able to stay alive in the climate, then its buds, roots and trunk have to be able to grow and it must be able to flower and set fruit, Finally, its seeds need to be able to germinate and survive. At the edges of a suitable climate zone, first the fruiting ability fails, then the flowering and finally the tree cannot survive.

For many trees, winter cold is essential and an average of 300 hours of cold (below 5°C) is needed in the UK to break dormancy and prepare for spring. Bud opening of beech trees is controlled by a combination of day length and temperature and mild winters can leave beech trees unprepared for growth, whereas hawthorn needs much less cold and is more ready to respond as soon as spring arrives.

Soil is a good insulator, keeping roots warm in winter. The bark around roots is less effective as an insulator than trunk bark and so roots are vulnerable to winter cold if they are exposed. Soil temperatures near the surface reflect changes in air temperature. At a depth of 1m (3ft), annual temperature changes are only a few degrees and at 2m (6ft) the soil temperature remains constant.

As temperatures rise, the difference in water vapour pressure increases between the air and the inside of the leaves. Unless there is an increase in humidity this leads to an increase in transpiration and tree water loss. An increase in temperature from 10°C to 20°C almost doubles the water vapour pressure.

Water

Water is the lifeblood of a tree. It courses around its body, having been drawn up from the soil around the roots, into the trunk and from there up to the branches and leaves. But trees do not have hearts to pump the water around; instead, they use the evaporation of water from leaves into the atmosphere – transpiration – to provide suction that pulls up continuous water columns the whole height of the tree.

In spring, deciduous trees use osmosis to help to get this process going – the high concentration of salt ions in the roots compared to the soil creates a pressure gradient that draws water into the roots. Some trees such as birch and maple generate quite a pressure of sap before the buds break and leaves open, as can be observed if their bark is wounded in late winter, when copious amounts of sap escape. But this pressure gradient is not sufficient to power the whole process; transpiration is needed for this.

The water drawn up through the roots and distributed around the whole tree includes nutrients and chemicals from the soil and starch molecules from photosynthesis. All cells need water to operate effectively. In general, water is drawn up through the xylem cells in the trunk and flows back down in the phloem cells. The phloem is directly underneath the bark, while the xylem is deeper within the trunk. So, the tree is acting like a fountain under very careful control.

Lack of water is the major limitation to growth for trees, given that they are growing within normal temperature variations. If trees do not have access to water resources within the soil, the primary supply will be from rainfall. Temperature and water availability are closely linked as transpiration and evaporation (evapotranspiration) rates increase with temperature and rainfall has a cooling effect on the air and soil in summer.

Wind

Wind is essential for trees. It brings rain; it powers transpiration from the leaves, which, in turn, draws up water and nutrients into the plant body; it stimulates growth of the trunk and branches by making them flex, as well as providing the transport for pollen and seeds of many species. But wind also shapes trees.

The drying effect of wind means that exposed leaves, buds and shoots can struggle to retain their moisture, leading to damage to the plant tissues and, ultimately, death. This can happen to a single leaf or bud, or it can happen to a whole tree, dependent upon the exposure, size and species of tree. But a dead bud can still be of benefit to its neighbour if it can lessen slightly the exposure of the next bud. Over

A wind-shaped tree.

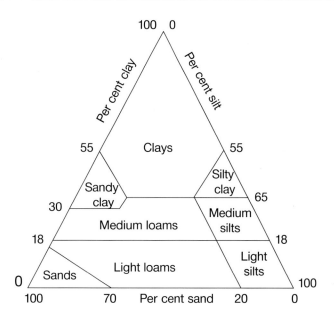

Soil texture triangle.

a whole tree, the buds growing in the lee of other branches or trees will remain healthier and continue to grow, downwind. So, over time, the resultant tree can look as if it is being blown downwind like a flag.

In general, tree roots grow in response to the stresses transferred down the stem to the base. Most trees are stable and able to cope with normal conditions. High winds will always be able to tip some trees over, but the risk of instability is greater when soils are wet, as this reduces their strength and the grip that the roots have on the ground.

Soil Properties and Nutrition

Soil is composed of soil particles, organic matter, micro-organisms, water and air. In a good soil, around 50 per cent of the volume is air. Natural soils develop over centuries through physical, chemical and biological processes acting on the underlying rocks, giving a mix of coarse and fine particles that form the mineral, non-biological, component of the soil.

Soil Texture

The soil particles give soil its texture: a soil composed of sand particles 0.2–2mm in diameter will feel grainy and gritty to the touch, whereas a soil with a high percentage of clay particles, of size up to 0.002mm diameter, will feel smooth and soapy. Silt particles are of intermediate size and have intermediate texture characteristics. So, in general, a grain of sand will be around 1mm in size, a silt particle will be one-hundredth to one-tenth of this size and a clay particle will be around one thousand times smaller than the sand grain.

Soil is classified according to the relative proportions of these different particles: a sandy soil contains less than 15 per cent silt and clay; clay contains over 40 per cent clay particles and less than 45 per cent sand and silt; a loam soil contains intermediate proportions of the three types of particles. Sandy soils are loose, well drained and well aerated, but can hold only relatively small amounts of water and minerals. Clay soils are denser and stickier, often poorly drained and aerated, but can store a large amount of minerals and water.

17

Soil Structure

Soil structure describes the characteristics of the soil. If the soil particles are the building blocks of its structure, the physical lumps that those blocks form when they are stuck together with organic matter dictate how the soil drains, how much air it can hold and how easy it is for roots to grow through it. In a well-structured soil, organic matter and root exudates glue soil particles together into lumps of variable size called peds. Soil structure can be affected by the management practices used on it; this is different from the soil texture, which is far harder to change.

Soil texture and structure together determine the size, amount and distribution of the pore spaces, which are the gaps between the particles and peds in the soil. In most soil, this pore space will take up 30–60 per cent of the soil volume. The smaller the volume of pore space, the more compact and dense the soil. In good soils, the weight per unit volume – the bulk density – is approximately 1.2–1.4g/cm^3 and root growth is generally suppressed where it is above 1.6g/cm^3, although there is some variation among different species.

Water in Soils

Between rainfall events, plants rely on the soil around them as a water-holding medium. During a heavy rainfall, water percolates into the 'macropores' (spaces between the peds) in the soil. The degree to which these are filled depends upon the intensity and duration of the rain and the drainage rate of the soil. When the rain stops, water drains out of these macropores, sucking air in as it leaves, and the amount of water a soil can hold on to against gravity is called its field capacity. This water can be sucked into plants by their roots.

Where rainfall is not frequent, the amount of water in a soil diminishes as vegetation uses it and roots have to suck harder to pull the water away from the soil particles. Once the pressure needed to do this exceeds –10bars (–10 times atmospheric pressure, or –1mPa), plants begin to wilt. At around –15bars (–1.5mPa), the water still present is too firmly held by the soil to be available to plants.

Nutrients

Trees need nutrients just as we need vitamins. Nutrients form vital parts of molecules in the plant body and perform very specific roles in a tree's metabolism. The vast majority of tree nutrients are taken up from the soil, although small amounts may be absorbed by the leaves from the air. While soils do not need to contain large amounts of the necessary nutrients, those soils lacking in nutrients will produce deficiency symptoms in the trees growing in them.

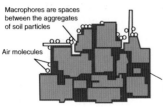

Water has filled all the spaces between the aggregates and is draining through the soil via gravity, unless drainage is impeded. As the water drains through it sucks air in behind it.

Macrophores are spaces between the aggregates of soil particles

Air molecules

Water molecules

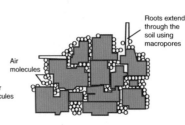

The water has drained away and now only that water that is held against the force of gravity is present. The soil is now at 'field capacity'.

Roots extend through the soil using macropores

Air molecules

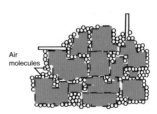

If soil water is not replenished, less and less will be left for roots to exploit. The force required by tree roots to prise the water away from the soil particles increases to the point where no more water is available.

Air molecules

Aggregates of soil particles, called peds or clods

Soil water.

WHAT'S THE DIFFERENCE BETWEEN ELEMENTS, MINERALS AND NUTRIENTS?

- An element is a chemical substance that can't be reduced to simpler substances by normal chemical means.
- A mineral is an inorganic element, such as potassium, sodium, calcium, iron, or zinc that is essential to the nutrition of plants. (There are only three organic elements: carbon, hydrogen and oxygen)
- A nutrient is a substance that provides nourishment for growth or metabolism. Plants absorb nutrients mainly from the soil in the form of minerals and other inorganic compounds.

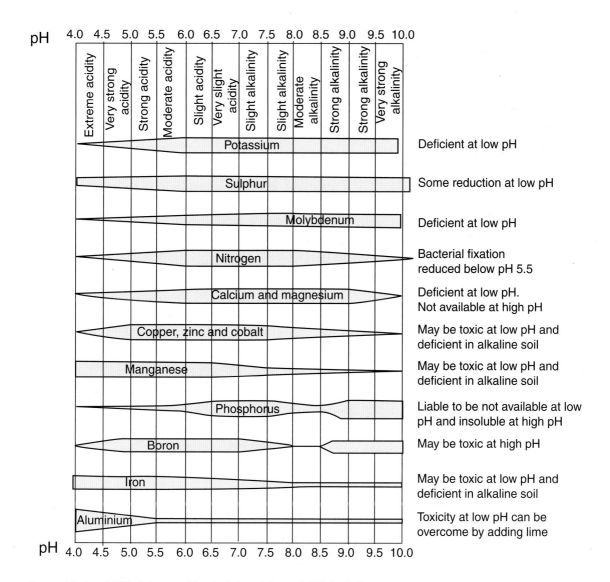

pH and nutrient availability. Thickness of bars indicates relative availability to plants. (Based on Roberts et al. 2006)

MINIMUM AND MAXIMUM pH RANGE FOR SOME UK TREES (COURTESY OF BARTLETT TREE SERVICES)

Species	pH	
	minimum	maximum
Alder	5.00	6.00
Apple	5.00	6.50
Ash	5.00	8.00
Aspen	4.00	5.50
Beech, European	5.00	6.50
Birch, Silver	4.50	6.00
Black Locust	6.00	7.50
Catalpa	6.00	8.00
Cedar (general)	5.00	6.50
Cedar, Western Red	5.50	6.50
Cherry (general)	5.00	7.00
Cypress	5.50	6.50
Cypress, Leyland	5.50	6.50
Fir, Douglas	6.00	7.00
Ginkgo	6.00	7.00
Hawthorn	5.50	6.50
Hazel	4.50	6.50
Holly, English	4.00	5.50
Hornbeam	6.00	8.00
Horse chestnut	5.00	6.00
Lime	4.50	7.50
Maple, Field	5.00	6.50
Maple, Japanese	6.00	8.00
Maple, Norway	4.00	7.50
Maple, Silver	4.50	6.50
Oak, English	6.00	8.00
Pear (general)	6.00	7.50
Pine, Scots	5.00	6.50
Plane tree	6.00	8.00
Plum, Flowering	6.00	7.50
Poplar (general)	6.00	7.00
Redwood/Giant Sequoia	5.00	6.50
Rowan	6.00	7.50
Sweet Gum	6.00	7.00
Sycamore	6.00	7.50
Yew	5.00	6.00
Tulip tree	6.00	7.00
Walnut	6.00	8.00
Willow (general)	6.00	8.00

Nutrients that are used in relatively large amounts (the macronutrients) are nitrogen, phosphorus, potassium, calcium, magnesium and sulphur. The micronutrients are used in relatively small amounts and are iron, manganese, boron, molybdenum, copper, zinc, chlorine and cobalt.

The elements that are present in the soil are released from the parent rocks by weathering and are usually in some form of chemical compound. A major by-product of weathering in moist, temperate soils is clay. Clay and organic matter determine the fertility of a soil by holding on to the nutrients in a way that makes them available to plant roots. Not all nutrients are present and available to plants in every soil and so some need the addition of fertilizer for trees to grow well.

pH

The soil pH is a measure of hydrogen ion concentration in the soil solution. Neutral soils have a pH of 7, with acid soils containing higher concentrations of hydrogen ions and lower pH. As pH increases above 7, hydrogen ion concentrations decrease and the soil becomes more alkaline. The pH scale ranges from 0 to 14, with most soils being in the range of 3–10.

The pH of a soil affects the availability of nutrients, with some elements reacting strongly with the hydrogen and producing compounds that are insoluble. Also, the amount of available aluminium, which is toxic to plants, increases with acidity.

THE BENEFITS OF TREES IN URBAN AREAS

In urban areas the environment is affected by the production of pollution, changes to the physical and chemical properties of the atmosphere and the covering of the soil surface. These cumulative effects lead to raised temperatures within the urban zone compared to the surrounding natural landscape. This is known as the 'urban heat-island effect'.

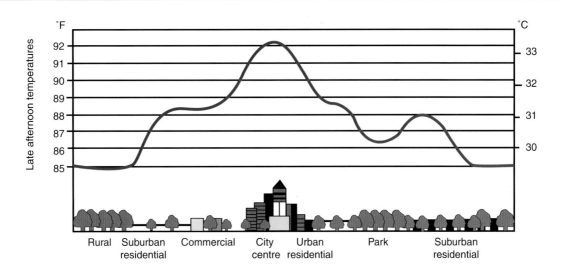

An urban heat-island profile. (From the EPA Heat Island Reduction Initiative)

Trees are part of the life-support system of our towns and cities and have an important role in countering the urban heat-island effect. They absorb carbon dioxide and fix the carbon into their woody structure. They also absorb air pollutants and fine dust particles, helping to renew and cleanse our urban atmosphere. During the day, they produce and release oxygen into their surroundings. Tree roots can act as sponges, absorbing water while their crowns intercept rain before it reaches the ground, which helps to reduce the effects of flash floods. Trees shade hard surfaces, thereby helping to cool them in summer and warm them in winter.

This is only the beginning of the benefits that urbanites receive from trees. Trees can screen roads, buildings, fences, walls and all other urban structures, or they can bring seclusion to an inner city open space. They provide a natural contrast to manmade structures, as well as adding scale to and framing their surroundings. As well as being a screen, trees can provide shelter and shade, either as individuals, or as a shelterbelt, avenue or line of trees.

Trees can be valuable components of slopes, where they add to the erosion control of vegetation by intercepting rain and softening its effects on bare earth. Tree roots by themselves do not provide significant erosion control, but as part of a plant community they can provide this service.

Trees represent natural features and they can act in concert with other vegetation to provide wildlife habitats, or they may do this job by themselves. Their naturalness is also valued as a seasonal clock, with trees broadcasting the time of year by displaying flowers, leaves, fruit or bare branches.

Urban communities value trees as being of aesthetic beauty and providing pleasant surroundings, or even maintaining contact with the history of an area. Trees can increase property values of an area and help to reduce the incidence of crime; deprived areas tend to be characterized by a lack of mature and growing trees, whereas safe, hospitable areas usually have a healthy tree population.

Trees in towns provide employment opportunities for many and working on trees produces products of economic value, such as timber, mulch, firewood, compost, charcoal or woodchip biofuel. Recent health research is beginning to quantify the many benefits of trees for reducing stress, promoting recovery from illness and drawing positive emotional responses from their scents.

Disbenefits of Trees

But trees do also have some disbenefits. They can damage buildings by extracting water from clay subsoils beneath foundations, or by physically pushing over freestanding walls, kerbs or other light structures. Tree roots exploit defects in drains and can contribute to their malfunctioning by forming a mat of fine roots inside a pipe. Trees may block important views, interfere with sight lines along roads or from CCTV cameras and they can create dark places where personal security becomes an issue.

Shade from trees has traditionally been seen in the UK as a disbenefit and in some situations trees definitely cast more shade on areas than is helpful. Debris falling from trees can cause damage or injury and become a hazard lying on the ground. Some things that fall from a tree can be even more unpleasant; pigeon and squirrel droppings may have human health consequences. Honeydew is a sticky secretion from aphids feeding on leaves, usually lime or maple. When this falls it is quickly colonized by yeasts that turn it black.

Trees can block television signals, although there is no automatic right to receive signals that justifies the removal of offending trees. The effect of trees on the atmosphere may not always be beneficial. Many trees produce gases besides oxygen. Volatile organic compounds are given off by growing trees and these can mix with other air pollutants to aggravate air pollution levels. But this is not a major side effect of trees; let's keep things in perspective.

REFERENCES

Kozlowski T.T., Kramer P.J. and Pallardy S.G., *The Physiological Ecology of Woody Plants* (Academic Press, 1991).

Park B.J., Tsunetsugu Y., Kasetani T., Kagawa T., Miyazaki Y., *The physiological effects of* Shinrin-yoku *(taking in the forest atmosphere or forest bathing): evidence from field experiments in 24 forests across Japan* (The Japanese Society for Hygiene, 2009).

Raven P.H, Evert R.F. and Eichhorn S.E., *The Biology of Plants*, 4th edn (Worth Publishers, 1986).

Roberts J., Jackson N. and Smith M., *Tree Roots in the Built Environment* (TSO, 2006).

Thomas P., *Trees: Their Natural History* (Cambridge University Press, 2000).

Ulrich J., 'View through a window may influence recovery from surgery', *Science*, 1984.

It is difficult to prevent pigeons perching on branches over paths.

chapter two

The Development of Towns

Most towns in England have developed from Anglo-Saxon villages. Market towns grew where farmers and merchants met, maybe at a river crossing or a port. Occasionally, whole new towns would be laid out and receive a charter from the king. These may have been a royal project or carried out by a baron or bishop, but, as often as not, they failed completely. New towns, or planned enlargement of existing ones, could also be the initiative of more local landowners, but the same limitations of funding and interplay of population and market conditions were crucial in determining success or failure.

Even as early as during the Roman settlement there were examples of forts becoming local hubs of commerce and government. These forts were typically of only 50 acres in size and the pattern was repeated in the early Middle Ages when Norman castles sprang up beside or within existing towns. Very few trees would thrive in a tightly packed castle or walled town of very limited size. But there would be garden space for vegetables and space to keep some animals, so it is not difficult to imagine fruit trees and an occasional shade tree. Towns that needed to be defendable, however, also needed to minimize obstructions to the views of defenders and anything of potential use to attackers. So it is unlikely that trees would have been tolerated close to town or castle walls.

It is also difficult to imagine townsfolk leaving trees that were in the way of an expansion of a market or inn, for instance. Conversely, no one would want the extra work of destroying a tree that was not in the way.

LIFE IN EARLY TOWNS AND VILLAGES

In Tudor times, a fairly large town would contain between 1,500 and 5,000 people. Most of the buildings were made of wood and mud, with drainage consisting of an open trench discharging into the street, where a central open sewer would remove the waste. Each householder was responsible for clearing the sewer passing their dwelling. The main streets would be lit during dark winter evenings by lanterns using candles or oil. The roads beyond the town were the responsibility of the parish council to maintain by paid or unpaid labour, so were of very variable quality. Road surfaces would be of cobbles or stones where these were available, but were usually just compacted soil, easily rutted and eroded.

Most people lived in the countryside and worked in agriculture. Hamlets and villages were less cramped than towns, but the houses and drainage arrangements would be similar. Trees in the countryside would naturally grow wherever there was a patch of ground left undisturbed by carts, people or animals. Wood was the primary fuel for cooking and heating, so trees would be favoured where they could be used in this way.

From before the Norman Conquest, much of the central part of England was farmed using the open field system. Wide, open spaces would be ploughed with oxen or horses, and long, straight runs with wide 'headlands' for turning the plough team were common. The land was parcelled up, with the main landowner given important areas of farmland and each villager receiving a narrow strip of good or poor land, as befitted his status. All the villagers would have 'common rights' to let their animals graze over the fields when crops were not growing there. This would help clear the land ready for the next crop and provide manure to fertilize the soil. Similarly, common rights would allow the villagers to collect wood from trees on the land or in woodlands nearby.

However, from Tudor times onward there was pressure exerted by landowners to enclose land for sheep grazing. This was increasingly profitable and required less manpower, but it eroded the common rights of villagers and so was resisted by local people. In the eighteenth century, enclosures were seen as necessary for improving agricultural efficiency and were encouraged by Acts of Parliament between 1750 and 1850. Over time, successive enclosures reduced the rights of the poor and made living in the countryside harder, which fed the growing movement of the workforce into towns. The process of enclosure led to the planting of hundreds of miles of hedging, in which many trees would eventually grow, and which has shaped the character of the countryside beyond Britain's towns.

Throughout the eighteenth century, technical developments improved agriculture and industry. For instance, the spinning jenny, introduced in 1764, greatly increased the capacity of each spinner to convert wool into long threads called yarn, but soon water power was replacing the local textile worker, leading to the development of large mill factories along streams and rivers. These, too, were overtaken by the technology of steam and the textile industry became large-scale and moved into the town, or towns developed round the mill factories. And where the factories were, people followed.

THE NINETEENTH CENTURY

By 1801, the population of Great Britain was 10.5 million. We know this because it was the year of the first national census, by which a specific programme of recording the numbers of the population was started. This census has been taken every decade since that time. The population figures hint at the gigantic changes that took place in the nineteenth and twentieth centuries. Whereas the population of England, Wales and Scotland increased by roughly 50 per cent every century up to the end of the eighteenth century, it doubled between 1801 and 1851, then doubled again up to 1911.

In 1801 just under 10 per cent of the population lived in London and 7 per cent lived in towns of between 20,000 and 100,000 inhabitants. By 1911, 80 per cent of the population was classed as urban. So while the population was growing fast, the growth of towns was increasing even faster.

However, life in a town in 1801 was not so different from that in a Tudor town. Buildings were more usually made of brick than timber, as this was less prone to catching fire and experiments were beginning with gas lighting. The usual method of dealing with waste was via cesspits beneath the floor designed to overspill into the streets, which might have a cobble or dirt surface, with a thick layer of rotting manure on top. In London there were 60,000 horses working in the streets by 1780, as well as thousands of other animals, and sixty sewers draining directly into the River Thames. Six of the nine water suppliers drew unfiltered water from the same river for city dwellers to drink!

Pall Mall in London had gas lighting in 1811, Westminster Bridge in 1813. By 1823 numerous towns in the UK had at least some of their streets lit by gas. There was enthusiastic uptake of this new technology, partly because gas lighting cost 75 per cent less than oil lamps or candles.

In 1811, London was around ten times the size of the next largest cities, Manchester/ Salford, Liverpool, Birmingham, Bristol and Leeds. Humphrey Repton, with John Nash, designed and built Regent's Park in London between 1812 and 1827, bringing the landscape of the country estate into the city. Some care was being taken to design and build gracious houses near the edges of London such as Chelsea, Belgravia and Bloomsbury, but no one was planning the housing of the thousands of workers drawn like a tide to the capital. Builders such as Thomas Cubitt concentrated on the top end of the market. His work emphasized broad and airy streets with spacious squares and formal designs that also included considerable numbers of trees.

By 1824 there were 24,000 power looms in production and the first railway from Stockton to Darlington opened in 1825. Michael Faraday invented the electricity generator in 1832, but no one yet guessed how important electricity would be to city life.

Cholera outbreaks occurred in England and continental Europe in the 1830s and 1840s. But still the connection between clean water and human health was not widely known. Most people feared the awful smell of sewage, but rather than getting rid of it the reaction was to seal the windows of houses at night against the 'foul air'. This led to many people dying from hydrogen sulphide gas or methane rising up from the cesspit under the floor.

People were being packed into towns, making homes in shoddily built houses and tenements. Basements and individual rooms were used to house whole families and often more than one family. No wash houses or toilets were provided; the only concern was commercial. In general, houses for the labouring class were

Gower Street, London. Built in the late eighteenth century, this aristocratic street is wide and spacious.

Roupell Street, London, built in the 1820s for artisans.

built by private builders. The buildings were not built to provide basic necessities; they were put up at the minimum cost to provide a return to the investor. For a long time there were no regulations to comply with and no standards to be met. This also meant no plan for the buildings in relation to each other, so blocks were built piecemeal.

A Royal Commission in 1844–5 reported that there was little that could be done in already crowded areas of large towns, but legislation was needed to regulate new street widths and to place drainage, paving, cleansing and water supply under a single administrative authority. This was a radical departure from existing practice and needed strong justification. The change was necessary, according to the Commission, to avoid severe problems in maintaining the labour force, restraining the working class and maintaining the rule of law. But there was a continuing and deep-seated resistance to legislation that impinged on private property. The Public Health Act 1848 and housing acts sponsored by Lord Shaftesbury in 1851 seemed to herald real progress in

controlling town growth, as a Central Board of Health was set up with powers over drains, water supply and cemeteries among other matters, but the same issues continued for the next twenty to thirty years.

During the 1840s, the railways leapt forward, connecting towns by thousands of miles of track. City stations were important buildings and much thought went into the architecture of these grand structures. The new railways brought more pollution, much material for industry and also thousands more workers into the towns. The commercial telegraph system grew with the railways, its principal customers in the first decades, as this provided an easy route for the telegraph poles, with few landowner issues to negotiate. Complications arose when access over private land was required.

In 1853–4, more than 10,000 Londoners died in cholera and typhus epidemics. This was an increasing problem, although it was not unique to London; all growing cities were vulnerable to such waves of disease. Sanitary problems in London came to a head with the Great Stink

---

A CENTURY OF ACCELERATING POPULATION GROWTH

- By 1811 the population of Great Britain was almost 12 million. London was home to 864,845 people.
- In 1821 the population just passed 14 million.
- In 1831 the population of Great Britain was 16.25 million. London contained 10.64% of the total with 5.71% in other towns with over 100,000 population and another 8.7% in towns down to a size of 20,000 inhabitants.
- By 1841 the population of Great Britain was 18.5 million and there were 1,873,676 people living in London.
- The population of Great Britain by 1851 was heading for 21 million.
- By 1861 the population of England, Scotland and Wales was over 23 million and there were 10 thousand miles of railway track in Britain.
- The 1871 population of Great Britain was 26 million and this had increased to nearly 30 million by 1881.
- By 1891 the population of the country was 33 million.
- At the start of the twentieth century, in 1901, the national population was 37 million.

---

in 1858. The summer brought an unendurable stench from the sewers, drains and the River Thames, which received the waste of the whole region. The smell was so bad that the Houses of Parliament considered closing down entirely. Something had to be done, quickly.

The improvements to the main sewers had been under way since the 1840s, but there was now increasing momentum to get things done. Chief engineer Joseph Bazalgette organized the design and building of a centralized sewer system that served most of London by 1866. This drainage system was more than the main sewers; it linked all the dwellings with small, 2in diameter tubular pipes connected to the larger, egg-shaped sewers and the main sewers, so that all the filth of the capital could be drawn away and treated outside the city. The water closet and flush toilet formed an integral part of this new sewer system.

Technological change and population growth made the introduction of relatively cheap railway or horse tram and omnibus transport practicable. In the 1860s, the first special, cheap-fare trains were provided for workmen and led directly to the massive building of working-class houses on the north-east edge of London served by the Great Eastern Railway. Where there were no cheap trains, the working classes crowded into those suburbs that had other cheap transport.

Electricity generators were being developed throughout the 1860s, but were not commercially successful until the Gramme dynamo was invented in 1870. The first electricity plant, at Holborn Viaduct in London, was not built until 1882. The Slum Clearance Acts of 1868 and 1875 and the Public Health Act 1875 finally brought progress in living conditions by regulating development through building by-laws. The Telephone Company was formed in 1878 to exploit Bell's patents in England, but progress was slow due to vested interests and complexities in expanding services between different suppliers and the government.

By 1881 there were forty-seven towns with populations above 50,000. Towns had to swell and engulf adjacent settlements wherever they could. The census figures over a number of decades show a sudden emergence of towns, packing in human multitudes as tightly and fast as possible, then, later, they show the need to spread people out again, in similar haste. For example, the population of the City of London, the central core of England's capital, shrank from 75,000 in 1871 to 50,500 in 1881, but the daytime population rose from 170,000 in 1866 to 261,000 by 1881. And it was the daytime population that used most of the city infrastructure, from the sewers to the roads to the new technologies. At the end of each day this throng withdrew to the outer suburbs to

Peabody Trust buildings in Clerkenwell, London. Built to improve living conditions of the poor in the 1880s – no room for trees.

regain the strength for another day's work in the city.

By 1890, London's suburbs were built up to the limit that could be served by horse-drawn transport. Other cities faced the same problems, but in London they were most acute. The situation was changed by the arrival of the electric tramway, followed by the electric railway and then by the appearance of the motor-bus. These modes of transport required a smooth surface, which provided the impetus for finding an effective way of making roads.

Bell's patents expired around 1891 and the trunk lines, owned by the National Telephone Company were nationalized in 1896. The company developed call offices round the country, which had become part of the government-owned Post Office by 1912.

By the end of the nineteenth century electricity was recognized as being more efficient for street lighting than gas and more and greater uses for this new power were being found. In the next two decades electricity use expanded massively, for example the Wolverhampton Corporation Electricity Lighting Extension Order in 1913 doubled the area it supplied with electricity from 5.5 square miles to 11 square miles.

THE TWENTIETH CENTURY

The national population in 1901 was 37 million, an increase of 4 million over the previous decade and by 1911 another 4 million had been added. Eighty per cent of the population of nearly 41 million was now classed as 'urban'. In 1909, the government passed the Housing, Town Planning, etc. Act. But it did little to change the situation in the short term. The bill only allowed control of new housing development schemes, so had no effect on existing areas of poor-quality housing. What it did do was to introduce the principle of town planning and this was gradually expanded through the next few decades.

During World War I, the government considered ways of unifying the electricity supply and the idea of a national grid developed, building and linking together power stations so that electricity would be available to everyone. Work on the national grid began in 1926.

The Housing, Town Planning, etc. Act of 1919 began large-scale state intervention to increase the supply of working-class houses by providing subsidies for housing. The new standard for working-class homes became a three-bedroom house with kitchen, bath and garden. These

Electricity pylon. The first electricity pylon was erected in 1927 near Edinburgh.

houses were built at a density of fewer than twelve per acre and to achieve this they had to be built at the edge of towns on open land. As a result council housing estates grew up alongside private suburbs. In the 1930s, 2.7 million council houses were built. The spread of urban areas into suburbs was fuelled by rapid development in transportation and this led to 'ribbon development' along major roads. Weak local planning authorities struggled to contain this growth; they were hesitant about blocking housing schemes because of the liability to pay compensation.

The size of Wolverhampton quadrupled to 47 square miles in 1925. The West Midlands Joint Electricity Authority was given responsibility to provide power to an area of around 1,000 square miles in central England. Industry took 80 per cent of this electricity and this proportion was likely to be typical for the whole country. In 1933, one-third of the 45 million population had electricity in their homes. The national grid was completed in 1937 and by 1944 electricity was in two-thirds of Britain's homes.

In 1939, at the outbreak of World War II, the estimated population in Britain was 46.5 million (there was some confusion of statistics at this time, due to mobilization of people for the war). During the war, work proceeded on how to cope with the continued growth of cities. The Barlow Report in 1941 was a major part of the post-war welfare state and the Abercrombie Plan of 1943–4 made explicit the need for open spaces within cities.

The war brought about centralized planning and control of population and industry due to the bombing of major urban centres. New factories were built and relocated in depressed areas, where there was significant unemployment. But as well as this shift to regional planning, the war brought opportunities to redesign the damaged cities. City planning became a practical necessity, not just a theoretical philosophy. The Barlow Report had set out the advantages of urban concentration as proximity to markets, reduction of transport costs and availability of labour. But the report considered that there were also serious disadvantages, including high land values, time lost through traffic congestion and inefficiencies from tiring commuter journeys for the workers. The report recommended the development of garden towns, satellite towns and trading estates to ease urban congestion.

The Town and Country Planning Act, 1947, brought together the thinking of policy-makers during the war and took into account the massive damage to infrastructure. As the population grew, so did the number of households, adding further to development pressures. Development of urban areas, and urbanization of other areas, was now firmly under government control and direction.

By 1946 London was home to around 20 per cent of the population and Britain was almost bankrupt due to six years of global war. Nationalization of many major industries

Milton Keynes showing typical wide verge giving more root space for trees.

followed. New towns sprang up as part of the rebuilding of the nation after the war in the 1950s and 1960s. Since that time the new towns have proved themselves. For example, the large-scale tree planting in Milton Keynes is now maturing and adds to the character of the town and Telford is a model of UK urban forestry management.

The post-war efforts to limit and control urban expansion were a world away from the nineteenth-century rush into cities. And yet that same old rush to development and narrow perspective of speculators persist. The Commission for Architecture and the Built Environment (CABE) in their 2009 report, *Who Should Build our Homes?*, lamented that the housing boom was primarily about chasing volumes and shareholder return and resulted in the majority of housing built in the UK not meeting the necessary standards. With the sudden end of that era following the financial crash in 2008, the national construction landscape has changed; or has it?

The essential elements of urban life are recognized and understood by a growing number of professionals, but the room for manoeuvre is always limited. 'Planning', in the sense of having a clear vision for the future and targets for achievement, struggles to become

a reality because of the existing sprawl and infrastructure, entrenched land interests and inertia in large, bureaucratic machinery. Long-term commitment is needed, which is rare in a culture of short-term political expediency. But given that this has always been a major obstacle, what has been achieved over the last 150 years is remarkable.

ROADS

From 1555 to 1835, each parish was responsible for the roads in its area. Initially, maintenance work was undertaken and paid for by local people. Gradually, though, more paid maintenance was used. However, this system was ineffective and the quality and extent of roads in the country was poor. A new approach to the problem was tried in the seventeenth century when justices were given direct responsibility for the upkeep of important local roads, with the power to collect tolls from road users.

From the start of the eighteenth century, private companies, or 'turnpike trusts', applied for permission to carry out this task. Most of these roads were around 20 miles (32km) long, but this varied and in Exeter the toll road network around the city stretched for almost

150 miles (240km). The initial idea was to return the road to local control after twenty-one years, by which time it should be in better condition. But usually a 'trust' would apply to renew the agreement. After 1750, trusts were required to install milestones along their roads to inform travellers of the distance between main towns.

The first act of a turnpike trust was to erect toll booths at strategic points along the controlled road. The road was blocked using a heavy barrier of pikes, or sharpened metal stakes, fixed to a horizontal bar. The toll collector usually lived in the toll house beside the barrier and collected the toll from users of the road before access was granted. Coaches, horses and other animals being led to market would be charged different rates. Long-distance coaches would pass through many sections of toll roads on a journey, but the improved road surface made it quicker and more comfortable and, potentially, safer from highwaymen.

The upkeep of the turnpike roads was a major responsibility of the trusts and many materials and methods of road construction were tried in an effort to produce hard-wearing and low-maintenance surfaces. Thomas Telford reduced the time taken for the London to Holyhead mail coach from 45 hours to 27 hours by sloping the road from the middle to either side, so that water shed away from the surface easily. John Metcalf in Yorkshire developed a method of construction which specified laying large stones as a sub-base, replacing excavated road material above this, then adding a layer of gravel to form the surface. John Loudon Macadam, another influential engineer, designed roads with broken stones laid carefully in a tight, symmetrical pattern and covered with small stones, creating a hard surface. These roads were called 'macadam roads'.

By 1825, there were 1,000 turnpike trusts controlling 18,000 miles (29,000km) of road. But this was just a small proportion of Britain's roads overall and the situation was about to change. With the coming of the railways, the importance of the turnpikes diminished rapidly.

As tolls collected dwindled so investment in the roads reduced and many trusts became bankrupt. From 1871, therefore, the Turnpike Trusts Commission arranged for existing acts to continue in force, but with the set objective of discharging debts so that the roads could pass back to the local authority, often a local highways board. The 1888 Local Government Act passed responsibility of main roads to county and county borough councils.

Up to the end of the nineteenth entury, roads in England were built in accordance with the prescriptions developed by Telford, Metcalf and Macadam, but all sorts of surfacing materials were experimented with, including timber planks, wooden blocks and even rubber. The first significant use of asphalt in England was in 1869.

Edgar Hooley combined tar and slag to form smooth roads at the start of the twentieth century, marketing his idea as 'tar-macadam'; the name 'tarmac' was born. However, as far back as the 1850s there were roads in Paris that were built using naturally occurring asphalt (a thick, sticky liquid or semi-solid) to bind the road stones together. In 1870, the Belgian Edward de Smedt, an immigrant to New York, patented a way of sealing road stones together using this material. As road construction rates increased, asphalt had to be synthesized. It was produced industrially as a by-product of the distillation of crude oil. In Britain the distillation of coal produced town gas that was piped to urban homes. The process also produced tar, similar to asphalt, which was used extensively as the road-binding agent in the first half of the twentieth century.

With the suburban expansion and building of new towns after World War II, the need for more and better roads intensified, as car ownership doubled in the 1950s, then doubled again in the 1960s. Road building reached deep into the countryside, linking towns and villages, but also carving wide, manmade corridors through communities and natural features. Today, the Highways Agency looks after all the trunk roads and motorways in England, a total of 4,818 miles

Trees soften and add interest to the Embankment, London.

(7,754km) in 2004, and traffic movement and congestion are major national issues. Traffic levels have increased in the United Kingdom by 25 per cent between 1994 and 2009.

TREES AS URBAN ELEMENTS

In the eighteenth century, trees in towns would be found primarily in private 'pleasure gardens' or on private property, mostly on the edge of the built-up area. Any tree that managed to survive surrounded by thousands of feet (animal and human), which led to compaction and pollution, as well as pruning, damage and interference, was rare and only tolerated if it was not obstructing business, or had cultural significance, such as a Gospel Oak or a hanging tree. As the nineteenth century got under way, new building for the upper classes included avenues, vistas and street trees, but this was localized and no one bothered about such things where tenements were thrown up to maximize profit.

As the Victorian period progressed, trees became recognized as valuable elements for inclusion along new roads. The Embankment in London was the result of the new sewer system and was planted with London planes in the 1860s. Growing towns expanded to engulf existing trees at the roadside, but nurserymen had the skill and wisdom to know how to plant trees beside existing streets. But as trees were being put in, they were also being removed; new street layouts did not have to preserve existing trees.

The road network, within and between towns, extended more and more through the twentieth century, with large trees becoming surrounded by concrete and tarmac, as if they were being engulfed in a rising urban tide, with nature retreating deeper into the countryside. Despite this, trees have always been planted in British towns. We are the most gardenized country in Europe. And nature has the capacity to use the most obscure pieces of neglected land, thus softening, fractionally, the hard, urban landscape.

Ebenezer Howard's *Garden Cities of Tomorrow* was first published in 1902 and, over time, changed the way that town planners and governments viewed urban growth. His first garden city, Letchworth , was built in 1903 and was followed by Hampstead Garden Suburb, 1908, and Welwyn Garden City in the 1920s. Howard's approach was to lay out areas for development in concentric circles with gardens and plants prominently positioned to soften the manmade elements.

The concept of new towns was taken up more definitely after World War II and trees became important elements in the urban mix of these new settlements. Although no new towns have been centrally planned since the 1960s, twelve Community Forests were launched in the 1990s and these now cover over half a million hectares and serve nearly half the population. In these areas, community involvement in tree planting and management is being encouraged.

Tree preservation orders, part of the new planning system developed after World War II, identify trees as important visual elements and allow for the protection of strategic specimens in order to maintain the character of a locality or to retain prominent features. This principle of tree protection has been with us now for over sixty years and research is providing information about the depth and breadth of the benefits trees provide in towns.

Trees in Towns II

The *Trees in Towns II* study, by Britt and Johnston, provides valuable insight into the state of our urban tree resource at the start of the twenty-first century. Surveys were carried out in 2004 of all English local authorities and information was collected according to different land uses, size of towns, types and size of tree in each of the counties and regions.

On average, in urban areas over the whole of England, there are 5.84 trees and shrubs per hectare. Mean canopy cover of urban areas in England is 8.2 per cent and ranges from an average 11.8 per cent in the South East to 4.0 per cent in the North East. In the category including formal and informal open spaces, the mean canopy cover is 15.9 per cent and in low-density residential areas it is 22.8 per cent. In industrial areas it is only 3.6 per cent. Back garden trees provide a significant part of the canopy cover in all regions. The South East has the highest densities of trees over 20m (66ft) tall at 4.9 per hectare. Most regions have fewer than two per hectare. Most trees surveyed were between 2.5–10m (8–33ft) tall.

Leyland cypress tree. This hybrid tree grows fast and, left unpruned, will become a very tall tree on favourable sites.

Two-thirds of the trees in towns are on private property, mainly gardens, or on less accessible public land such as schools, churchyards, allotments and so on. Almost 20 per cent of trees are in local parks or on open spaces and 12 per cent are street or highways trees. Less than a one-third (31 per cent) of the urban tree stock is made up of large broad-leaved species of tree. Small broad-leaved trees contribute 42 per cent of the total and the remaining 27 per cent are conifers.

The number of tree and shrub species recorded varies from 148 in the North East to 196 in the South West. Species richness was greatest in low- and medium-density residential areas and least in industrial areas. This is understandable, as the number of species able to survive harsh, industrial conditions is bound to be fewer than those able to find niches within the mosaic of conditions found in residential areas. Also, the climate is milder in the South West than in the North East.

The total numbers of species need to be considered in light of there being only 35–40 native species of tree, but, also, many of these species are likely to be shrubs. The most frequently recorded species is the Leyland cypress (not a native but a hybrid of two cypresses from North America), which is twice as common as the number two species, hawthorn. Then come, in order, sycamore, silver birch and common ash. Leyland cypress densities are greatest in the South East. Sycamore is the most recorded large, broad-leaved tree and is most common in the North. Birch is most common in Yorkshire and the Humber, but also in industrial areas, probably as a consequence of it germinating from seed easily on open areas. Old ash trees form a relatively high proportion of trees in the North East. Privet is recorded as most common in London and the South East.

Two-thirds of trees are either in the young or semi-mature category. This is superficially encouraging, as there should always be more young trees than old trees, so that, over time, as trees are affected by planned and unplanned changes, there are sufficient numbers remaining to recruit into the next size and age category. But the results do show a worrying trend, in that the greatest numbers of trees are in the ten- to fifty-year age range, reflecting a surge of planting in the 1980s and 1990s following Dutch elm disease and two bad storms. The figures show that the impetus to plant has faded.

Recent Initiatives

As trees come under increasing pressure in urban areas their numbers dwindle, but at the same time the value of them to local communities becomes highlighted. With community support and ongoing research in the UK a groundswell of interest in trees as important elements of urban life has arisen, ranging across the spectrum from concerned tree professionals to other, allied professionals and to the people living in towns and cities who see trees every day and recognize their value. Trees are now acknowledged as being vital parts of the urban life-support system.

The Coalition government has announced a commitment to a national tree planting campaign. The government wishes to be the greenest government ever and to empower local people and urban communities to shape their own, local landscapes. The goal is to plant over 1 million trees during the next five years, with a significant proportion of these being street trees. Local groups, authorities and civil society organizations are encouraged to work together on focused, local initiatives, within the framework of a national partnership of government and non-government organizations.

Sycamore or Celtic maple. This common, native tree is most frequent in the north of England, but can be found in most areas.

MOST RECORDED TREES IN ENGLAND IN ORDER OF FREQUENCY (BRITT AND JOHNSTON, 2008)	
1. Leyland cypress	7. English oak
2. Hawthorn	8. Apple
3. Sycamore	9. Japanese cherry
4. Silver birch	10. Holly
5. Common ash	11. Rowan
6. Lawson cypress	12. Beech

In addition, the Mayor of London has launched an initiative to plant 10,000 trees by 2012 in forty areas across London that would benefit most from social, economic and environmental improvement. Trees for Cities started life in 1993 as Trees for London when a group of young Londoners recognized the need for more trees in the capital. The charity was established 'to advance the education of the public in the appreciation of trees and their amenity value, and in furtherance of this the planting and protection of trees everywhere and in particular inner city areas'. In 2003, the name was changed to Trees for Cities, largely in response to requests for support and advice from cities around the world. In addition to tree planting, the charity is now involved in a wide range of activities, including educational work with schools and community groups, vocational training in arboriculture and horticulture, re-landscaping the public realm, campaigning and fundraising.

The Tree Design and Action Group (TDAG) is a multidisciplinary group of individual professionals and organizations, from both the private and public sectors, working within the London Tree and Woodland Framework. The group has an interest in all trees in the urban environment and plans to publish guidance notes to increase awareness of the role of trees in the built environment that will have relevance in all urban areas.

The National Tree Safety Group is a broad partnership of organizations working together to develop a nationally recognized approach to tree safety management and to provide guidance that is proportionate to the actual risks from trees. Its membership includes professional bodies, tree owners and managers and heritage/conservation organizations.

REFERENCES

Ashworth W., *The Genesis of Modern British Town Planning* (Routledge & Kegan Paul, 1954).

Britt C. and Johnston M., *Trees in Towns II: A New Survey of Urban Trees in England and Their Condition and Management* (Department for Communities and Local Government, 2008).

Cullingworth B. And Nadin V., *Town and Country Planning in the UK*, 13th Edition (Routledge, 2002).

Hoskins W.G., *The Making of the English Landscape* (Morrison & Gibb, London, 1955).

NTSG, *Bringing Common Sense to Tree Management* (NTSG, 2010).

Raven P.H, Evert R.F. and Eichhorn S.E., *The Biology of Plants*, 4th edn (Worth Publishers, 1986).

Thomas P., *Trees: Their Natural History* (Cambridge University Press, 2000).

OVERLEAF: Central Milton Keynes, showing a wide boulevard planted with trees.

chapter three

The Urban Tree Interface

A dying oak in a hotel car park. This tree is probably suffering from damage caused during construction of the car park.

In urban situations, the conditions in which trees grow can vary from a woodland that is very similar to a rural area, to a roadside where a tree is growing in very limited, poor-quality soil characterized more by brick rubble and tarmac than by natural material. The average urban tree is likely to be enduring conditions closer to the latter than the former.

Not only are the soil conditions of wide variation, but the way the environment affects trees also is significantly different from in natural areas. Cities are warmer than the surrounding countryside, street lights, industrial and residential lighting alter the pattern of light and dark, as can reflective surfaces such as glass and metal. Buildings can have strange effects on wind. The urban pressures on a tree mean that it has to cope with high levels of stress resulting from the existing soil conditions and from exposure to a modified climate, with manmade effects introducing a damaging cycle of disturbance.

DAMAGING URBAN CONDITIONS

Light

Light is vital for photosynthesis, but it is also needed to direct plant growth and development. Trees that are adapted for growing in open spaces with good light levels will display greater stem elongation, smaller leaves and less branching when they are transplanted into shady locations.

In urban areas, light is present all day; when it gets dark, electricity is used to provide artificial light to allow work to continue and people to remain safe. But the levels of lighting in towns go far beyond what is necessary and artificial lighting is just another part of the urban environment that trees have to cope with. The annual cycles of growth and reproduction of many trees are controlled by day length, but can be altered by supplemental night lighting. Trees grow in response to three factors relating to light: the colour, or wavelength of the light; its brightness; and its duration. Photosynthesis requires quite high light intensities (around 1,000 microEinsteins/sq m/sec for light-demanding trees and around 200 mE/sq m/sec for shade-tolerant species), but only a fraction of these levels is needed to cause other, developmental changes in some trees. For example, a full moon seems to provide not enough light to trigger developmental responses in trees, but indoor lighting that allows reading, for instance, is 1,000 times more intense than this (4.6 mE/ sq m/sec) and a 150W fluorescent cool white bulb provides 17 mE/sq m/sec at a distance of 1.5m (5ft). Street lights can therefore fool trees into thinking that summer is still here, with long day length, although most deciduous trees eventually realize it is autumn and drop their leaves.

Daylight length can also influence leaf shape, surface hairiness, pigment formation, leaf fall in autumn and root development. It can affect the onset and breaking of bud dormancy. Where night lighting alters the natural photoperiod experienced by trees, it can have an impact on all these developmental processes. Trees that are sensitive to day length can be affected by artificial lighting after dark; for example, some plane trees in Bristol remain in leaf the whole year, apparently due to the presence of adjacent lighting. The continued growth is vulnerable to damage because it is not

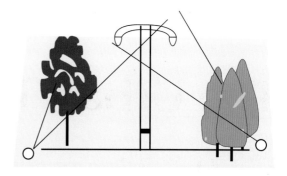

Poor lighting design. Unshielded lamp and upward-directed spotlights. Even proper selection of lamp type to minimize direct effects on trees will not prevent wasteful night sky pollution. (From Chaney, 2002)

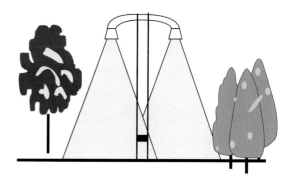

Good lighting design. Appropriate choice of lamp type provides night light but minimizes light pollution and effects on trees. (From Chaney, 2002)

responding in a coordinated way to other environmental factors, such as coldness. Younger trees can be affected more than older trees due to their greater vigour. Continuous lighting is more damaging than lights that are turned off late in the evening.

Artificial lights do not all have the same effects on trees. High-pressure sodium lights are the most damaging, as they produce light in the red to infrared wavelengths. Incandescent filament lamps are older types of lighting, but they emit a wide spectrum of wavelengths that

THE SENSITIVITY OF TREES TO ARTIFICIAL LIGHT
(ABBREVIATED FROM CHANEY, 2002)

High	Intermediate	Low
Norway maple	Japanese maple	Beech
River birch	Red maple	Ginkgo
Silver birch	Honeylocust	Sweet gum
Tulip tree	Red oak	Black pine
False acacia	Small-leaf lime	Pin oak

includes red and infrared. Where these lights are used, sensitive tree species should not be planted.

Where the choice of artificial light can be influenced by tree physiology low-pressure sodium lamps or mercury vapour lights should be recommended, as these do not emit light in the red to infrared wavelengths. Metal halide or mercury vapour lamps are also recommended, as they have limited light emission in the red to infrared regions. Lights should also be specified to be directed toward the ground to avoid 'spilling' the light in directions where it does no good and can affect trees. Turning lights off during off-peak hours is the main way to prevent disruption to normal plant growth. It is important to use the minimum amount of light needed to do the job and to use properly shielded lamps.

A plane tree in London, before a building is constructed on the south side.

Although additional lighting can increase photosynthesis, it is not always the case that more light means more photosynthesis. Too much light damages the chlorophyll within leaves and disrupts photosynthesis. Chlorophyll is most effective in moderate light conditions. Leaves that form in bright sunlight have to grow in a way that protects the chlorophyll from this deterioration. So, artificial light, by increasing light levels within leaves beyond a moderate range, can lead to less effective working of the chlorophyll within a leaf.

Where trees grow close to large, glass-sided buildings, additional light may be shone on the leaves by reflected sunlight. Short periods of this light during the day are not likely to damage leaves except where the temperature rise is very marked. Normally shaded leaves exposed to bright light can suffer tissue damage, raised transpiration rates and damage to the chlorophyll, with the result that photosynthesis can be affected and the risk of infection raised. Conversely, a mature tree can be affected by being enclosed by tall buildings. The lowering of the light levels will not be noticed quickly, if at all, but the branch structure and general health of a tree could be affected over a period of years.

Temperature

Low Temperature Damage

Urban areas do not change the basic response of trees to temperature, but the concentration of structures and heat sources means that towns are some degrees Celsius warmer than rural areas. The most vulnerable parts of trees to low temperatures are the leaves and flowers. These may be damaged by chilling, frost or freezing conditions. Chilling damage occurs in temperatures above freezing when a plant is designed for milder conditions, such as sub-

tropical plants growing in temperate climates. The surface tissues are damaged and killed, which can lead to loss of growth and infection.

Freezing can disrupt water uptake and exacerbate existing damage to bark, branches and roots through the action of freeze-thaw. The main purpose of bark is to protect the wood within from the extremes of the outside environment. So, trees are vulnerable to both high and low temperatures where the bark has been damaged.

Water is present within the cells of trees in two main locations: inside the cells themselves and between the cells. The amount of water inside the cells depends on the osmotic concentration of salts in the cell, the cell sap concentration. The more water a tree holds inside its cells, the greater is its resistance to frost damage. This is because the cell water is protected by a plasma membrane and a higher concentration of salts that turn the water into an anti-freeze solution.

Same plane tree with the building nearing completion.

So, in freezing conditions the water between the cells turns to ice before the water inside the cells. But as this happens, more water is drawn out of the cells and the cell protoplasm contracts. The damage caused is dependent on how well the tree is able to recover from this 'plasmolysis', and the loss of water causes more problems than the formation of ice.

The amount of frost damage to trees is affected by the rate of freezing and thawing. If freezing happens very quickly, even frost-hardy plants can be affected. Rapid thawing means that large amounts of water are released in and between the cells and this can flood the plant tissues, making them unable to function. Young shoots exposed to early morning sun are especially prone to this damage.

New shoots and expanding buds are particularly vulnerable to frost damage in early spring when temperatures can plummet quickly below freezing. When this happens, shoots, leaves and blossom can be killed. The concentration of solid structures in urban areas can create artificial 'frost pockets', in which cold air (which is heavier than warm air) collects. This makes conditions harder for trees in these areas. In some cases, a clear line of damage can be seen in the lower crown of trees showing the depth of cold air. Obviously, small, young trees can be completely immersed in these 'pools' of freezing air and therefore damaged severely.

When freezing conditions occur in early autumn, they can catch trees out. Where trees have not fully ripened their growth, or 'battened down their hatches' in preparation for winter, cold temperatures can cause damage that is often not noticed until the next spring. This kind of injury to the cambium can damage the production of new growth of xylem and phloem and affect the main branches and trunks. This effect can also occur if there is a very mild period within the winter followed by a sharp drop in temperature. Some trees may begin to 'de-harden' and are then more vulnerable to the reappearance of winter weather.

The ability of trees to cope with cold temperatures during winter is affected by the rate of temperature fall in autumn, drainage, natural protection, species of tree, character of the root system and the specific combinations of conditions experienced. Tree roots are generally more sensitive to cold temperatures than the rest of the plant body, although they are normally insulated from the cold by the surrounding soil. However, where soils are poorly drained and the water freezes, roots can be damaged. Lack of undergrowth adds to the vulnerability of roots.

Trees that have been encouraged to continue to grow into autumn by nitrogen fertilization of the surroundings, for instance on golf courses or playing fields, can be injured when cold weather strikes. However, some trees will harden off in response to the changes in day length regardless of other conditions.

Heat Injury

Damage caused by heat and flames – scorching – is distinct from that caused by liquid or vapour, which is known as scalding. Most damage to trees from heat is a form of scorching according to this definition. Even sunscald could be classed as a scorch, as it does not need wet conditions to occur.

Sunscald typically occurs over a period in winter when surface temperatures can vary quickly between freezing and heating from sunlight. On cold winter days in the northern hemisphere the sun can increase the temperature of the south side of a stem by 20–30°C. Not only does this dry out the stem, but, when the sun sets or is obscured, the temperature drops rapidly and freezing of the tissues can result. Usually this sunscald damage occurs on young trees that have been recently planted. It can also occur on older trees that are suddenly exposed to greater intensities of sunlight, for instance where a group of trees is heavily thinned out, or where a tree crown is significantly thinned. Previous injuries to a tree increase the likelihood of sunscald.

Sunburn is a form of scorching caused by high temperatures induced by sunlight or hot winds during the growing season. This damage is more easily seen on leaves, fruit and flowers, as these parts of trees are less insulated, but it can also affect young or suddenly exposed stems. The affected parts dry out and cell membranes are damaged. The desiccating effect is the primary trigger for wilting and leaf death.

Extremes of heat are produced by fire and lightning. Both these factors are present in nature, in urban and non-urban areas, but fire is a particular, added hazard for urban trees. As one Australian fire-fighter pointed out, 'there are three main causes of fire: men, women and children'. These three causes are abundant in urban areas. The high temperatures dry out and damage the tissues of all parts of the tree and also affect the surrounding soil. The effect of lightning on trees is variable. Trees can be blown apart, the sap can instantaneously boil and they can burst into flames , or they can be scarred, or maybe just a single branch killed. Often the root system, or a section of it, is killed.

In urban or urban-related activities high temperatures associated with dry-heating processes (such as road surfacing) and wet-heating processes (such as venting of steam from industrial or air conditioning pipes) can occur. The results are either scorching or scalding of plant tissues.

Soil

Building in towns and cities often requires reshaping of the terrain by infilling, changing the gradient of slopes and removal, transfer and spreading of soil materials. Repeated demolition, refurbishment and rebuilding over decades, possibly centuries, means that urban soils often include large volumes of rubble, wood, glass, plastic, metal and chemicals.

Urban soils are characterized by a surface layer of mixed, filled and contaminated soil. Typically, development involves scraping away topsoil, shaping the subsoil and burying waste material such as rubble before other subsoil is added. Finally, topsoil is respread using either the original or imported material. Such soil

A fire-damaged tree.

usually has very poor structure. The result is often a very unnatural soil profile with sharp contrasts in texture, structure, organic matter content, pH, bulk density, aeration, water-holding capacity and fertility.

Compaction

Vehicles and pedestrians tracking back and forth over soils, especially wet soils, cause compaction, which greatly decreases the number and size of the macropores and so reduces root extension. As bulk density increases, roots struggle to push past the soil particles and squeeze through the squashed soil spaces. Compacted soils remain wet and undrained in wet conditions and become hard in dry spells because they have no obvious structure; the peds have been crushed and the macropores squashed, so there is little water available to the roots. Soil compaction does not

An urban tree under pressure.
(Photo courtesy of Jonathan Fulcher)

kill trees quickly, but, over a number of years, it can seriously weaken trees and make them vulnerable to other problems.

Soil compaction may be limited in intensity and distribution over a site, maybe to immediately around a footpath or road, but where there is widespread vehicle traffic over a protracted period compaction can be very extensive. Where compaction occurs around large trees it can lead to instability. Grassy areas may have a relatively open soil structure near the surface, but may be heavily compacted deeper down. In urban areas, bulk density can be as high as 2.0g per cubic cm. Roots growing in compacted soils tend to be shorter and thicker than those in uncompacted ones.

Nutrient Deficiencies

Urban soils may be deficient in nutrients at the time of tree planting, or they may develop deficiencies later. Soils in towns can be extremely variable, but, unless it's obvious that there has been no disturbance, it is reasonable to assume poor soil quality and nutrient availability. Much land in urban centres has been developed and redeveloped over centuries, with the soil now containing a very unnatural mix of brick dust,

cement, tarmac, glass, metal and timber. Over decades, soil can come into being from the breakdown of vegetation and animal matter, but it is a very slow process. Existing trees cope with these conditions by ranging far and wide with their root systems, gleaning any nutrients that are available from nearby better soils, or where nutrients are deposited by water or air action.

Deficiencies may develop over time if all the debris falling from trees is routinely collected and carted offsite. The leaves, twigs and other material falling from trees is an important part of the nutrient cycle. The tree absorbs nutrients from the soil and builds the plant body. Leaves contain those elements that cannot be drawn back to the twigs and these nutrients eventually fall to the ground. If they are removed, the cycle is broken. Over time, the cycle becomes a one-way process of mining nutrients from the site. What's left is soil of poorer and poorer quality.

Toxicity

Several substances can occur in toxic concentrations in urban soils. When this happens, it is an important issue, but it is not that common and is usually limited to specific parts of particular sites. For example, heavy metals may be present, but although these will contaminate the soil, toxic levels are uncommon, being restricted to sites associated with industrial or mining activity. Methane damage can be encountered, for instance in soils above landfill sites, or where gas leaks from pipes. Methane itself is not toxic, but it can quickly kill trees by driving out the oxygen and reacting to produce carbon dioxide, thereby asphyxiating them. On industrial processing works and gas works chemical compounds such as phenols may be found, but this is fairly rare.

Where salts accumulate to high concentrations, this can seriously inhibit tree growth. The main effect is in inhibiting water absorption due to the high osmotic pressure of the soil. But the high concentration of ions also has toxic effects. Some tree species are more tolerant of high salt content soil than others as their root systems can prevent it from reaching the shoots, but continued exposure to very high salt levels is likely to be damaging to all trees.

Water

In urban areas, water availability has a significant effect on trees. Large areas of hard surfacing, rain-shadow caused by tall buildings and the erratic effects of large-scale drainage and leaky water pipes can all lead to uncertain supplies. Urban trees are therefore vulnerable to both drought and waterlogging.

In forests, research has shown that the amount of water available accounts for 70–80 per cent of tree diameter growth. There is no equivalent data for urban trees, but the same principle is likely to apply. Solitary trees need around 800ltr (180gal) of water per square metre of crown per year for unlimited growth, or 600ltr (130gal) per tree for a line of trees. Of this figure, 40 per cent is considered to be the absolute minimum for vital growth (see Konijnendijk et al.).

Drought

Imagine a rain storm in a city. The raindrops start as occasional spots on the pavement, then build up in quantity and noise into an event where thousands of litres of water fall onto hard surfaces, such as concrete, tarmac, walls or roofs and, although puddles form and sink into exposed soil or cracks in surfacing, it does not take long for the drains to lead away the vast majority of the water.

Waterlogging caused by changed drainage.

When it rains in the city centre, up to 100 per cent of the water runs off the ground surface and into drains. For industrial areas, the proportion of ground sealed beneath hard surfaces may be 90 per cent and for family houses it is around 50 per cent. In parks, cemeteries and other open spaces the situation is better; only about 15 per cent of the water runs off without infiltrating the soil. This drain-off leads to a chronic drying out of soils and associated problems for trees growing in them.

Flood and Waterlogging

The opposite problem is also common. Drainage channels can lead water to impermeable areas, maybe because of compaction or blockage by underground structures, and water can remain in the soil with no way of passing through it. Storm drains can become overloaded, leading to surface water flooding. In these cases, tree roots suffer from waterlogging.

This same problem can result where groundwater levels are subject to large fluctuations due to drains or underground water movement. Roots that are frequently submerged in water in soil that does not drain quickly are very liable to the build-up of toxins in their cells and to asphyxiation due to lack of oxygen.

Poor drainage can have a similar effect to compaction, because the result is that roots cannot grow through the water-filled macropores and air fails to circulate for the same reason, limiting energy for root growth. This can lead to a distorted rooting pattern and even unstable trees.

Where soil is waterlogged there is no oxygen available to the submerged roots and they have to respire anaerobically. The by-products of this process are toxic to most trees and can only be tolerated for short periods before the roots become damaged. This is especially a problem in the growing season. Trees are more vulnerable to this in the spring and early summer, but there is some variation between species.

Where run-off is directed into a tree planting pit, the difference in compaction between the pit and the surrounding soil can have the effect of a submerged impermeable bowl and the pit will then fill up with water. This can happen regardless of the quality of the soil in the tree pit. Poorly drained sites limit the depth of the rooting of trees and often this leads to the development of shallow rootplates and unstable trees. Site drainage, therefore, is crucial in establishing and managing trees.

Wind

In Britain most urban areas are at low altitude, but this does not stop damage to trees from wind. Wind damages trees by scorching the leaves, drying out buds and wearing down leaves and shoots. It raises levels of transpiration and so increases water usage. In newly planted trees this can cause problems, as the root system is not large or connected to a large water supply in the surrounding soil. New trees can also be rocked in their planting pit if not planted correctly. This prevents new root growth and exacerbates drought problems.

In maritime areas, young trees are exposed to salt spray that can further damage leaves and reduce growth. Salt spray is not only restricted to coastal areas; salt-laden winds may be experienced up to 40km (25 miles) inland. Where wind is known to be a factor, it may be worthwhile considering planting large groups of trees rather than individual specimens. Groups provide shelter and support for each other and the outer stems may be expendable once the central trees have become established. However, watch out that you do not just expose trees in turn to damaging winds by removal of the edge tree. If this situation develops, just keep as much vegetation on the site as possible to form a windbreak for more sheltered trees.

In urban areas, the surrounding buildings can alter natural air flows and result in wind-tunnel effects. Trees planted in the way of these wind flows are especially likely to become desiccated and misshapen and it is easier to move the tree than redesign the buildings. Large trees suffer from wind damage during storms, or from

THE BEAUFORT SCALE: SPECIFICATIONS AND EQUIVALENT SPEEDS FOR USE ON LAND

Force	Equivalent speed 10m above ground		Description	Specifications for use on land
	miles/hour	knots		
0	0–1	0–1	Calm	Calm; smoke rises vertically.
1	1–3	1–3	Light air	Direction of wind shown by smoke drift, but not by wind vanes.
2	4–7	4–6	Light breeze	Wind felt on face; leaves rustle; ordinary vanes moved by wind.
3	8–12	7–10	Gentle breeze	Leaves and small twigs in constant motion; wind extends light flag.
4	13–18	11–16	Moderate breeze	Raises dust and loose paper; small branches are moved.
5	19–24	17–21	Fresh breeze	Small trees in leaf begin to sway; crested wavelets form on inland waters.
6	25–31	22–27	Strong breeze	Large branches in motion; whistling heard in telegraph wires; umbrellas used with difficulty.
7	32–38	28–33	Near gale	Whole trees in motion; inconvenience felt when walking against the wind.
8	39–46	34–40	Gale	Breaks twigs off trees; generally impedes progress.
9	47–54	41–47	Severe gale	Slight structural damage occurs (chimney-pots and slates removed).
10	55–63	48–55	Storm	Seldom experienced inland; trees uprooted; considerable structural damage occurs.
11	64–72	56–63	Violent storm	Very rarely experienced; accompanied by wide-spread damage.
12	73–83	64–71	Hurricane	

carrying excessive additional weight such as ice and snow or smothering plants. Wind action on these branches can flex them beyond their strength.

There is no way of completely preventing tree damage from high winds. The Beaufort scale is a subjective measure of wind speed that has been in use for nearly 200 years. It is calibrated for use on land by observing the damage a wind causes to standing trees. Therefore, by definition, high winds will have the capacity to damage trees; it is a visual indication that the safety factor built into their growth is being tested. Much of the damage caused to trees by wind is not just a function of wind speed. A tree is designed with branches that sway in the wind and this action allows it to dissipate the force acting upon it. In many cases, the damage is caused when gusts of wind follow each other and continue to blow against a branch that has no opportunity to return to its state of rest, so it bends, then bends again, then bends some more, until finally it breaks.

Pollution

Trees differ from humans in experiencing two distinct environments at the same time. We breathe the air and are affected by its changing chemical composition as we inhale and exhale, with our bodies reacting and responding accordingly. Trees also are subject to air as the leaves respire and transpire. But trees also have an important interface with the soil and below-ground conditions. Here, the same cause-and-effect mechanism operates, with the whole plant being affected by the soil constituents, nutrients and contaminants surrounding the root system.

Pollution can be airborne, as a result of gases and particles poisoning the air, or soil-borne, from chemicals leaking onto and into land. If a chemical lands on the ground, it can affect roots directly or by modifying the soil and if it lands on the trees it can interfere with biological processes. Pollution is deposited either in a dry form, as gases come into contact with a surface, or as a wet deposition carried in rain or snow, known as 'acid rain'. Deposition by cloud droplets usually has more concentrated pollutants within it than rain and is a particular problem over high ground, or where mists and fogs are prevalent.

Chemicals contaminating the air affect plant growth by physically damaging leaves and reducing their effectiveness in photosynthesis. This affects the supply of carbohydrates and other growth-regulating molecules, and therefore can inhibit both shoot and root growth. Polluting chemicals in the soil can injure roots directly and reduce their efficiency in absorbing water and nutrients, or can affect soil pH and availability of nutrients.

Pollution can be classified into three basic types: natural toxic substances found in unusually high concentrations because of human activity; toxic substances that are only found as a result of human activity; and substances that are not toxic themselves, but that combine with other substances in the environment to cause damaging effects. In regard to urban tree populations, pollution is visible as foliage damage, or poor growth and death. Pollution types one and two are toxic from their originating source and are referred to as primary pollutants. Secondary pollutants, type three, are those produced by interactions between primary pollutants.

Derelict land does not necessarily imply contamination, even though it is likely to be highly disturbed and mixed with foreign material. However, soil contamination in urban situations is very common and a survey carried out by the Department of the Environment (DoE) in 1993 defined derelict land as being that which is so damaged by development that treatment is necessary before it can be brought into positive use.

Air Pollution

The major primary air pollutants are sulphur dioxide, fluorides and particulates. Major secondary pollutants are ozone, nitrogen oxides, volatile organic compounds (VOCs) and peroxyacetylnitrate (PAN). Often these pollutants are carried onto plants or soil by rain, snow or mist, but they can also be deposited in dry conditions. Primary pollutants can react in the atmosphere to form other damaging compounds.

Non-polluting rainfall should have a pH of around 4.6. Over large regions of the world rain and snow can be much more acidic than this. Individual storms can be hundreds of times more acidic than normal.

Most secondary pollutants are formed by the reaction of fuel exhaust gases, such as nitrogen oxides, hydrocarbons and carbon monoxide, with the atmosphere in the presence of sunlight. Ozone, nitrogen oxides, VOCs and PAN are this type of pollutant and are called photochemical oxidants.

- **Primary pollutants** Sulphur dioxide is a by-product of coal burning. Levels in the atmosphere have fallen sharply over the last twenty years. Sulphur dioxide dissolves in water to form acids (sulphuric and sulphurous acids) that can acidify soils and damage leaf surfaces. The legacy of high sulphur dioxide levels is very acidic soils in urban areas. This leads to lack of microbial activity, lack of available nitrogen and a low carbon to nitrogen ratio.

  Fluorides are formed in the manufacture of aluminium, steel and phosphates. They are used in the production of glass, bricks and ceramics. They are the most phytotoxic air pollutants and cause damage to plants at very low concentration levels. However, fluorides are not found as pollutants as often as sulphur dioxide, ozone and nitrogen oxides.

  Particulates are tiny pieces of solid material, such as dust, cement, smoke and aerosols, that are held in suspension in the air. Their effect on vegetation depends on their chemical composition and how they may be absorbed into leaf surfaces. They can be carriers of toxic chemicals, as well as physically blocking the light from reaching leaves, or interrupting the workings of the stomata and gaseous exchange. Dust is normally a problem only near to quarrying, landfills, or where traffic levels are high.

- **Secondary pollutants** Ozone is the most important photochemical oxidant. In the higher levels of the atmosphere ozone is beneficial, shielding the Earth from ultraviolet rays. In the lower atmosphere it is a damaging chemical that can reduce shoot growth and the height of trees. Ozone

levels are often low in urban centres and higher in suburban neighbourhoods. It can be transported long distances in the atmosphere, with levels varying with the season and time of day.

Nitrogen oxides (NO_x) are highly reactive gases. The levels of NO_x have risen over recent decades, because levels are related to the amount of vehicle exhaust gases. NO_x can be seen as a reddish-brown layer in the air in urban areas when it is combined with particles in the air. Nitrogen oxides mix with water readily to form acid rain.

Air pollution tends to reduce plant growth by inhibiting leaf formation and expansion, damaging leaves and hastening leaf fall (abscission). It reduces height growth, cambial growth and flower and fruit growth. But pollution can also deliver nutrients to trees, especially nitrogen, which may be a significant benefit for trees in highly constrained, urban conditions.

The effect of pollution on trees depends on the type of pollution (and interactions between pollutants), concentration of the pollution, the duration of the pollution episode and on the environmental conditions before and after the episode. Trees cope with pollution either by avoiding uptake of damaging chemicals, or tolerance of such chemicals within the plant body or within biochemical processes. Trees are at greater risk of damage when very young or very old and tolerance of pollution varies with species.

Road Salt

Road salt is used by highways authorities to disrupt areas of ice on roads and to improve the grip of vehicles in icy conditions. In the UK, the main materials used for this purpose are sodium chloride ($NaCl_2$) and calcium chloride ($CaCl_2$). The build-up of these chemicals in tree root zones affects water uptake, soil pH and structure, nutrient availability and leaf health.

In coastal areas, sodium and chloride ions are found naturally in the air as salt spray that is deposited on soil and vegetation. This effect can be noticed far inland, but at low concentrations is not harmful to trees. In exposed areas close to the sea, concentrations can be very high and can affect tree growth.

The use of road salt on roads, paths and pedestrian areas is a problem for trees, because as it is washed into the root zone it affects the levels of salt in the soil. This interrupts the process of osmosis whereby water is drawn from the soil, a place of low salt levels, up into the tree, where the concentration of salts is higher. The difference between these salt levels is called the soil water potential and the greater the soil water potential, the greater the uptake of water. So, where road salt raises the level of salts in the soil it reduces the soil water potential gradient, thereby slowing and limiting water uptake by trees.

The chloride in road salt can become a problem for plant growth and injury to leaves is associated with chloride concentrations of above 1 per cent in broad-leaves and 0.5 per cent in conifers. The salt also reduces, prevents or destroys rooting in the surface soil layers, which can be a problem particularly for surface-rooting species.

Sodium ions can be very damaging in soil. A soil that is accumulating salts over time will gradually become more alkaline. In clay soils, when sodium ions saturate the clay particles the structure of the soil collapses, reducing aeration and root penetration. The sodium displaces other nutrients needed by the trees and seems to reduce the activity of soil micro-organisms and earthworms.

Road salt has the same effect as natural sea spray on the soil and vegetation. The effects are particularly noticeable where road salt is washed off roads around trees and where snow is dumped and piled up near to trees. Fast-moving traffic tends to lift the salt into the air and can cause more damage from salt spray than from run-off. Trees growing in depressions suffer greater damage than trees on raised sites and trees on downhill sides of roads also suffer more than trees on the uphill sides. Damage seems to be limited to within 30m (100ft) of roads, with most severe damage being within 5m (16ft) of a highway.

Spring thaws and rains tend to wash away some of the salts and allow some recovery to take place. A survey carried out in London in 1990–1 found that quickly condemning plane trees after winter road salt damage was premature and that it is advisable to monitor affected trees in late summer before any drastic decisions are taken.

A salt bin, showing damage to adjacent vegetation. (Photo courtesy of Jo Ryan)

Heavy Metals

Common heavy metals include copper, mercury, lead, nickel and zinc. These elements can damage animal and human health and also affect tree growth. Heavy metal contamination occurs as a result of nearby mining, or other industrial activity on or near to a site. Heavy metals, such as copper, zinc and nickel, are particularly associated with traffic vehicles.

Petrol fuel used to be produced with a high lead content and vehicle-combustion processes resulted in a high deposition of lead onto surrounding land. This situation has changed since the adoption of lead-free fuel in 1986, but lead can stay in the soil for decades at least.

Busy roads can be contaminated by zinc, which makes up 2 per cent of vehicle tyres. Build-up of sodium levels from road salt in soil can result in the release of heavy metal ions from clay particles, which can then get washed into watercourses.

Contaminated sites will be recognized either by knowledge of their history or previous studies and documents, or their presence may be noticed by the pattern of vegetation growth. Testing for heavy metal contamination that is toxic to trees is a complex process requiring the input of laboratory techniques and specialists. The presence of heavy metal ions within the soil may be very high, but if these ions are strongly fixed to the soil there may be little effect on tree growth, and measuring metal concentrations in leached water may not identify the level of contamination effectively.

Plants require some metals for growth. Molybdenum, manganese and zinc are needed at very low concentrations, as is boron (a non-metallic element). Aluminium is present universally in soils and can become toxic if the pH is too low.

Trees help to control contamination in a number of ways. They increase water use on the site and so reduce leaching of the heavy metals into nearby watercourses. They reduce the effects of wind blowing away contaminated soil particles. They also take up heavy metals into the plant body and store them there.

However, if the metals end up in the foliage, this fixing will be for only a short time and the leaves can be blown off the site and contaminate nearby land.

Different species store different elements in different parts of the plant body, so it is important to look carefully into the tolerance of tree species and the way in which they achieve this before planning to use trees on a contaminated site.

Soils contaminated with heavy metals can be treated in a number of ways:

- adding lime to the soil raises the pH, which tends to lock the heavy metals in position on the soil particles, reducing leaching; the solubility of the heavy metal ions increases with low pH
- soil can be shipped off-site and replaced with good-quality soil, but this is expensive
- the use of tolerant tree species can help to revegetate the land.

Spillages and Fire

Hydrocarbon compounds can be produced by burning materials on a site, or by spillages of fuels and oils. Such activities are common on development sites, or within industrial sites and the effects can last for decades.

Once soil is contaminated in this way, the remedial options are limited. Removal and replacement of the soil is a drastic remedy, but may, ultimately, be necessary. In the USA, vacuuming of contaminated soil and its replacement by good-quality topsoil is recommended if existing trees are to be retained. Where young trees are to be established, smaller volumes of soil can be replaced, but then the trees will be left to deal with the wider soil contamination once their root systems have grown.

Carbon Dioxide, Methane and Ammonia

Landfill gases, including methane, carbon dioxide and monoxide, ethylene, ethane, hydrogen sulphide and propylene, are produced as rubbish decomposes. Methane and carbon dioxide make up over 90 per cent of these gases. Carbon dioxide and methane kill roots immediately by displacing oxygen within the soil, thereby effectively asphyxiating trees. Carbon dioxide is also toxic to tree roots. Escape of landfill gases into soil produces very similar effects in trees to waterlogging.

It seems that the best trees to use in such circumstances are those that have shallow root systems, in which oxygen is less likely to be depleted. Trees that can tolerate waterlogging would also be worth considering for planting in such situations. It has been found that young, newly planted trees are more able to cope with such damaging conditions than existing or larger trees, probably due to their small root systems being able to avoid areas of high gas concentration. Bear in mind, though, that shallow-rooting species are vulnerable to drought conditions.

Ammonia is produced in many biological processes and is typically found in high concentrations on or near animal farms or enterprises. It is also a by-product of incineration, coal burning, sewage treatment and large-scale fertilizer application. Ammonia acidifies soil because of its high hydrogen content and also because it improves the efficiency of sulphur dioxide absorption into the soil.

SPACE AND DISTURBANCE

Urban conditions impact on trees in the way that the space around them is used and abused by man. Development pressures mean that space in towns and cities is used ever more intensively, with the result that building and infrastructure step closer and closer to the trees. This leads to the removal of trees that are in poor condition and those that are compromised by the new tree–building relationship.

The trees that are left become subject to more stringent constraints, as the natural growth of roots is blocked by foundations or soil compaction and branches are cut back when they abrade roofs, gutters or walls and affect light levels through windows.

Farther from buildings, trees can still be affected by excavation, leading to root pruning, tearing and crushing, for underground services, including drains and soakaways. Changes in levels are frequent in gardens as owners 'improve' the site by making it more level or introduce retaining walls to form terraces. Reducing soil levels exposes roots, which are then usually pruned and removed from the site, whereas raising soil levels buries roots and can result in their asphyxiation.

Above ground level, as infrastructure approaches the trees, more people are brought into close contact with them. This leads to vehicle collisions breaching the bark insulation of the trunk, or pruning is required to keep roads and paths clear. Trees can cope with being used as noticeboards, posts for hanging wires or Christmas decorations, even for all the paraphernalia relating to tree houses, but only if natural growth is taken into account and the trees are checked periodically and fixtures removed, repositioned or adjusted. One hole in the trunk is not necessarily a major problem for a tree, but many holes, repeatedly driven in deep, or larger and larger ones added over time, can overwhelm the tree's capacity to cope.

Trees, as living things, have similarities to human beings; a bit of stress now and then can be a nuisance, but also moulds and shapes them, and life continues. Continued stress and damage, and stress coming from an increasing number of threats, will, over time, grind down resistance and resilience, leading to weakness and vulnerability to pests and diseases. Trees are beautiful natural features best appreciated when they are thriving and allowed to grow well.

A tree damaged in a vehicle collision (probably a mower).

REFERENCES

Bradshaw A.D., Hunt B. and Walmsley T., *Trees in the Urban Landscape* (E. & F.N. Spon, 1995).

Chaney W.R., *Does Night Lighting Harm Trees?* (Purdue University, 2002).

Hibberd B.G. (Ed.), *Urban Forestry Practice* (Forestry Commisson, 1989).

Konijnendijk C., Nilsson K., Randrup T.B. and Schipperijn J., *Urban Forests and Trees* (Springer, 2005)

Kozlowski T.T., Kramer P.J., Pallardy S.G., *The Physiological Ecology of Woody Plants*, (Academic Press, 1991).

Roberts J., Jackson N., Smith M., *Tree Roots in the Built Environment* (TSO, 2006).

Roman A., Cinzano P., Giacometti G.M., Giulini P., *Light Pollution and Possible Effects on Higher Plants*, Societa Astronomica Italiana (provided by the NASA Astrophysics Data System, 2000).

Thomas P., *Trees: Their Natural History* (Cambridge University Press, 2000).

Urban J., *Up By Roots* (ISA, 2008).

chapter four

Planning Urban Tree Management

BASIC PRINCIPLES

The essence of urban tree management is to grow and control the tree resource in a limited area in order to optimize benefits to the owner or users. In this phrase is embodied the idea that trees require periodic inputs to maximize their benefits and to minimize disbenefits and risks to people and property. It also recognizes that trees are, generally, positive features on a site, a view endorsed by the *Trees in Towns II* study, which found that 96 per cent of urban trees were looked upon as making some contribution to the urban environment. This view is also supported by research in the UK and the USA, which indicates a high value placed on 'urban nature' across diverse cultures and communities (*see* Schroeder *et al.*, 2006).

Tree management begins before a tree is planted, covers the time when the tree is becoming accustomed to its new environment (establishment phase) and extends through its formative period, as it grows to full size and becomes a mature component of its surroundings. Management continues into the period when the problems associated with the tree accumulate and decisions have to be made about removal and replacement.

The management process operates at different levels. At a strategic level, it includes planning the future contribution of trees to an area and deciding between competing objectives and constraints. At the day-to-day level, it informs immediate and short-term decisions that need to be made about trees. This process may operate over a wide urban area, or it may be contained within a single road or garden. Regardless of size, the same general considerations and planning are needed.

Planning is the basis on which any progress is made and is underpinned by information, from existing resources and knowledge and from specific surveys and assessments.

PLANNING TOOLS

At the smallest scale, where only one tree, one garden or one small area is the focus of the management, planning begins with a piece of paper and a pencil. A plan of the site is always of value, as it helps to visualize the space available. Objectives for the area need to be clarified and this then informs decisions about what species to plant to get the right effect within the desired timescale. So, here we see many of the issues that need to be addressed over wider areas focused on one point. Information needed before planting is carried out includes:

- type and characteristics of the soil and climate

- purpose of the planting
- timescale over which any tree is expected to be a part of the landscape.

In any specific situation, from planning a small garden to managing a city's urban forest, the same questions need to be considered if long-term benefits from the trees are the goal:

- What is the long-term purpose of the tree planting?
- Do the trees have a single role, or do they meet a number of objectives for the local community or the city?
- How tall, wide, thick and dense should the trees become?
- Will deciduous trees fit the bill, or are evergreen trees required?
- If screening is an objective, will breaking up the outline of the eyesore be enough, or does it need to be completely obscured?
- Do the trees need to be able to screen the eyesore immediately, or is it acceptable for them to grow up to do this job?
- What is the prevailing wind direction?
- Is the site prone to flood, contamination, drought or frost?
- Is the soil sandy or a clay, or compacted?
- What characteristics of the site might influence tree species and planting positions?
- What are the condition and value of the existing trees and surrounding landscape?
- Will neighbours object if trees are planted close to their boundaries?
- What grounds maintenance work is needed around the new trees to encourage quick establishment?
- What protection will the newly planted trees need?
- Are the trees in danger from children, animals or vehicles?

These factors will either indicate appropriate species that are likely to thrive, or they may clarify that changes to the site are needed before any planting is attempted.

Are small trees wanted, or fast-growing ones that will become large features in future? Speed of growth is an advantage when a tree is planted that needs to attain a given height, but it becomes a disadvantage if the tree continues growing, becoming a hazard to others, or requiring frequent pruning to keep it to the right size. Trees are less traumatized and grow fastest when they are planted at small sizes. Larger trees need more careful aftercare and may take longer to become used to their new surroundings and to put on height and spread. However, if cost is no object ready grown trees can be planted to perform the given role immediately.

Are there other, secondary roles that the trees could perform? For example, providing fruit, encouraging wildlife, shade, decorative flowers, fruit or foliage? Maybe there is a distinctive character to the surrounding area that is worth enhancing. Most situations have a number of factors that inform the management choices and the best success is achieved when they are all properly considered at the outset.

Don't forget to consider the disbenefits of particular tree species, such as dense shade to the south of a window, sharp spines, poisonous fruit, large cones that could injure people, or large, squashy fruit that could be a slip hazard on a footpath. Any of these characteristics could make tree species inappropriate or a liability if a choice is made to plant it without careful thought.

Finally, issues of cost need to be thought about. Not just the cost of buying and planting the trees and checking their quality, but other factors like:

- What will it cost to modify the site by decompacting the soil or improving the drainage?
- What aftercare will the trees need?
- What formative pruning will be needed in the future?
- Will there be a need to prune the trees regularly and are there specialist contractors in the area?

Urban trees and woods are often well used and valued by the local community, but without spending time finding out how the resource is used it is easy to overlook or miss part of its role in the locality.

So, with as much of this information as possible being known, the choice of species can be made confidently and work can begin on any necessary modifications to the site to ensure the new trees will thrive and money won't be wasted.

As the size of area increases, so does the number of factors to consider and the range of management options. At a strategic urban forest level planning is an interdisciplinary activity, drawing together many diverse objectives. To omit the planning element from a site of whatever size is to condemn the tree resource to uncontrolled actions that are likely to undermine its effectiveness to provide specific contributions to the surroundings.

Local authority tree officers are likely to be involved at either the planning and development control level, or through the day-to-day management of publicly owned tree resources. In each case, there is a further dichotomy between daily management and future planning. Some things need to be taken slowly, making sure that the policy framework or an agreed set of principles are in place before an activity commences. But, equally, a lot of things just won't wait for someone 'upstairs' to decide; a decision and an action are needed today.

In England today there are only 16 per cent of local authorities that have carried out a full survey of their highways trees. Only 19 per cent know what percentage of their district is tree-covered and only 28 per cent have an existing, specific tree strategy (Britt and Johnston, 2008). These are very low levels of information to be working from. The picture when it comes to strategic planning is slightly better, with just over half of all local authorities having some type of strategy referring to trees. But these bare facts just highlight that where there are planning strategies that include trees they are not often based on comprehensive data about the tree resource.

Trees in Towns II looked into the most relevant district-wide strategies and found that 42 per cent of them did not include any

provision for the revision of the plan and 75 per cent did not include any targets for tree planting and management, but focused on broad policy objectives.

The simple basis of the planning process is that, to be effective, it must be an ongoing process, continuously open to new data and changes in values. The basic planning model asks three questions and includes a feedback loop.

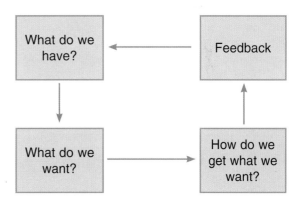

Basic planning model.

Van Wassenaer has a similar approach, which extends beyond deciding how to get what we want into the carrying out of plans and operations to reach the set goals. Life is guaranteed not to be as straightforward as our plan, so adjustments and reassessments will be needed.

What Do We Have?

Before you can begin to formulate a plan for the future, you need to know what you have. Baseline surveys collect this information and analysis of the results gives an insight into the present situation, as well as probably what's happened in the recent past. Even at this initial

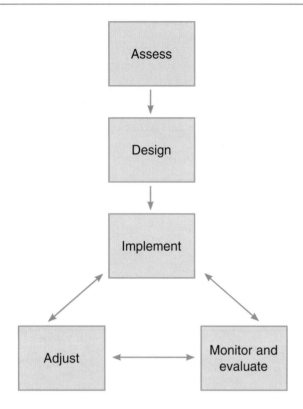

An adjusted planning model.

stage of the planning process, the feedback loop is needed. You need information on which to base plans, but to get that information you need to think about what data needs to be collected.

> You have to know what you've got before you can manage it.

Baseline studies dealing with tree resources will collect information such as tree numbers, tree species, condition, age class and size parameters. Land use categories may be employed to show different uses and opportunities for tree planting and management. Of course, many agencies or

authorities will have some information available already, so an early step in this process should be to check what is in the files and what is already known.

At a strategic level, other data is needed about planning in a wider perspective. The value of the tree resource may be realized by comparisons with studies carried out in other areas. Other professionals may have information that relates to trees, such as parks strategies or expected population growth, current recreational land allocation, plans for future industrial or community development and so forth. Establishing some value of a tree resource should be an early objective, as showing how valuable a tree population is and what benefits accrue to a community from it is an important step in identifying and liberating funds for its management.

What Do We Want?

Formulating overall goals can be a difficult process, as all the people who have an interest in the tree resource – the stakeholders – should have some input into the debate. But it is a vital element; leaving out any segments of the community could fatally undermine the effectiveness of a management plan. Conflicting objectives are likely from different stakeholders. At the strategic level, this could involve discussions with development planners, local communities and specialist departments (such as highways or housing).

At the day-to-day level, it may mean gaining an understanding of the tree resource, tree ages and condition and the prevailing climate and ecology. At this stage, the focus is on understanding the situation and finding areas of agreement. When dealing with conflicting objectives or views it is important to work at achieving a consensus between involved parties. Compromise is often necessary to move forward.

Although urban trees are not recognized sufficiently as being of high value, except when very large or part of a local cultural perspective,

Clear objectives for management will make it easier to know when you are making progress, how fast you are moving and when you have achieved success.

their contribution to the urban life-support system will grow over the coming decades and become a vital element in keeping our cities inhabitable.

How Do We Get What We Want?

The process of bringing together relevant information should, when allied to a careful assessment of the trees, help to identify appropriate management objectives. These objectives then inform the management activities that are carried out either directly on the tree resource or at the strategic level to prepare the way for the development of the resource. Although in Britain we have nearly 200 years of experience of living in large towns and cities, we have not often included trees in our strategic planning. Deciding on an objective is, therefore, likely to need some creative thinking and investigation of available technologies before ways can be found to make the vision a reality.

Feedback

The planning process is an iterative process, which means you don't just do it once, but rather the steps are repeated as more information becomes available. However, this assumes some monitoring element is being carried out. Reassessment is necessary periodically to check if values have changed and if the management operations are resulting in the expected benefits.

Computer Programs

The collection, analysis and synthesis of information is greatly helped by use of a computer. However, a paper-based system may be sufficient for a small site containing few trees. Such a system would organize the information and keep records of principles and objectives of management and necessary operations with dates of their action and results. The system should include some feedback and review, but does not need to be complicated.

Where the tree resource is more extensive, with greater numbers of trees and overlaps with other professions and a range of diverse stakeholders, computers are needed. First, the computer helps to keep up to date with the changing policy environment. Valuable documents can be gleaned in electronic format that feed into assessment and analysis of a situation. Overlaying of map-based information from diverse sources can prove invaluable, but usually takes time to achieve. Google Earth satellite photos are now being used to assess tree canopy cover over large areas and it is possible to cross-reference site plans showing trees with this resource and so add to the precision of estimates, as well as providing more confidence in identifying specific trees for those carrying out site visits.

Landsat imagery can also be used for mapping canopy cover over large areas, but the pixel size of around $30m^2$ ($100ft^2$) may be too coarse for urban forest planning. High-resolution imagery is more detailed, giving information down to a $1m^2$ ($3.3ft^2$) pixel size. Both these options are relatively expensive and involve inaccuracies of up to 15 per cent, whereas photo-interpretation using Google Earth or other aerial photographs can be a lot cheaper and purports to be 97 per cent accurate.

At ground level, software is available that makes data collection straightforward and streamlined. Tree surveys usually take the form of collecting specific size and condition parameters for trees, street trees, parks trees or some other distinct, identifiable population. Generally, the more information collected, the slower and more expensive is the process. Conversely, the more generic the information that is collected, the more limited is the analysis that can be carried out. The lesson here is to think ahead. What will the information be used for? If the information is only intended to draw up a maintenance schedule, then tree species, location and necessary work may be all that's needed. However, if the plan is to build a case that the tree resource is of exceptional value, it helps to know other parameters, such as crown size, height and diameter, so that some value of the environmental services provided can be computed. Maybe there is a need to track the spread of disease through the tree population; in this case, additional information about symptoms and pathogens is needed.

There are several software systems commercially available in the UK, including Arbortrack, Confirm and Ezytreev. These packages combine a powerful server or office computer with a hand-held data collection device, such as a Husky, or a tablet notebook such as a Panasonic Toughbook. The latest mobile phones are becoming powerful hand-held computer devices and it is expected that these will be used for data collection in future. Only 56 per cent of local authorities have computerized tree-management or inventory systems (Britt and Johnston, 2008). This means that there is plenty of room for improvements with councils introducing these tools for the first time.

SITE EVALUATION

On receiving responsibility for management of a site, the first requirement is for information. Management decisions must be based on an informed position if subsequent actions and operations are to be successful, so the first objective is to reach that informed place. Once the information is available, options can be considered and appropriate objectives set. Obviously, the information may reveal gaps in knowledge and there will always be a need to go on learning about a site.

Robinson in *The Planting Design Handbook* (2nd edn) sets out a helpful way of approaching surveying a site, which is useful in showing what needs to be considered during any evaluation or assessment process. He separates this work into four main areas:

- physical survey
- biological survey
- human survey and
- visual survey.

It will not be necessary or practical to carry out all of these surveys and assessments for every site, but it is important to be familiar with the range of information that can be collected when evaluating urban sites.

Physical survey

Geology

The geology of a site may not be immediately apparent, but it will influence the natural soil on the site, which will in turn affect the trees that can be successfully grown there. There are good geological maps of the UK that can be used to provide this information. If a site does show a significant presence of underlying rock, it is a sign that there is not a lot of soil present. The British Geological Survey has a series of geological maps, at 1:50,000 scale, that provide comprehensive information about the nature, extent and age of the different rocks within a district. Only around half the country is covered by these maps, but there are also regional maps available. These *British Regional Geology Guides* are a series of paperback books that offer an overview of the UK's geology. All these maps and resources are marketed by Ordnance Survey.

Topography

The orientation of a site can alter the way the climate affects it, and also, therefore, the way that trees respond when planted there. Slopes that face south will receive more sunlight than a comparable site facing north. This added sun will dry out the site more and so a north-facing site will tend to be damper and cooler than a south-facing one. Also, slopes facing a prevailing wind will be more exposed and subject to storm damage.

Cold air flows, as it is denser and heavier than warm air. This means that cold air at the top of a slope has a tendency to roll down the hill. Any obstruction to this movement can lead to a build-up of cold air, for instance around buildings, a wall across the slope, or a natural feature such as a wood or belt of trees. This could be important if trees are grown in these 'frost pockets', where they are more likely to get damaged by the cold.

Altitude needs to be considered on any site, as it may be a limiting factor for some plant growth. However, most urban sites will be within a known, limited range of altitudes in which urban foresters and arboriculturalists are familiar.

The slope of a site is an important natural feature. It may produce a boggy bottom and an arid scarp, or it may gently angle the land up to the southern sun. Drainage is strongly affected by slope, but drainage systems, natural springs and a significant build-up of structures on a site can change this natural pattern of water movement.

Climate

Temperature is a fundamental factor in determining whether a particular species will grow well on a site. Temperatures that are too high will result in scorching and wilting of the tree, while temperatures that are too low will lead to frost and cold damage. In either situation plants will not grow well. It is important to be clear about these minimum and maximum temperatures, as they can affect trees even if they occur in quite short durations.

Rainfall is a primary factor in influencing the growing conditions on a site. The expected total annual rainfall and the way it is spread throughout the year are valuable pieces of information to help decide what is likely to grow well.

Meteorological information can usually be found for an area quite simply. The Meteorological Office Rainfall and Evaporation Calculation System (MORECS) is a nationwide service giving real-time assessments of rainfall, evaporation and soil moisture over 40km square grids throughout the UK. However, the Meteorological Office website includes ongoing weather data from UK weather stations that is compiled monthly and summarized annually. There is also easy access to historical weather station data for many sites in the UK. These provide valuable sets of climate data tracing back many years of maximum temperature, minimum temperature, air frost days, rainfall and sunshine hours. For instance, the Hurn weather station has data stretching back to 1959; the Ross-on-Wye station data goes back to 1930 and the Bradford station records started in 1908. However, a note of caution is needed, as the information is limited to specific weather stations that may not be representative of the site being considered. The Meteorological Office website also warns that the relevance of urban observations to the surrounding urban area can be difficult to judge, as some of these observations are made on rooftops. For any long-term planning of large areas, it makes sense to consider setting up a weather station to collect directly relevant data.

Wind affects tree growth physically by blowing branches away from the prevailing wind direction and also by drying out leaves and buds. Trees 'exposed' to strong winds are moulded by its effects. This is seen most clearly in 'flag' trees growing on mountain slopes. Here the trees become misshapen due to extreme wind speeds during winter. In urban areas, and by the coast, similar, but less extreme effects can be seen. In towns this is usually the effect of increased exposure to wind and low temperatures due to the funnelling effect of buildings.

The prevailing wind direction will have a general effect on tree growth and this could also result in increased vulnerability to storm damage. Such damage is usually not the effect of wind alone, but may be combined with an increased weight on a branch or tree due to snow or ivy for instance.

Alfred Place in London is a typical, busy street. This London plane tree is surrounded by concrete, tarmac and vehicles. Do the occupants of the adjacent building value its shade and softening effect on the urban view or do they see it as a dominating presence that shuts out the sun and increases maintenance costs for the building?

Microclimate

Microclimate refers to the conditions experienced on the ground, in quite small areas. We have already mentioned frost pockets, but in an urban situation great variation of soils can be encountered in just a few square metres. Wind funnelling can mean that careful positioning of new trees is necessary to avoid planting a tree that will inevitably get blown to pieces or dried to a husk.

Recording existing plants can give an indication of soil and environmental conditions and particular attention should be given to changes in the flora that could well indicate a change of microclimate conditions. Obviously, a survey may not be focusing on microclimate details if it is a city-wide effort, but a comprehensive garden survey may need to include this information. Information gathered should be relevant to the scale of survey and clearly specified at the outset.

Hydrology and Drainage

If drainage plans are available, use them. On a site, drainage, or its lack, may be indicated by changes to vegetation. Boggy ground needs to be recorded as such. Information about local springs can be very helpful when formulating management and planting plans.

Existing Structures and Services

Any site assessment must take into account what is present at the time. Buildings obviously need to be considered, but so do paths and roads. Often, on derelict sites building foundations may be left after the structures have been removed. These are still a constraint on the site and should be recorded. Topographical surveys pick up these features as a matter of course and, if they are available, they are valuable plans on which to base any site assessment.

Check with records offices and council departments to see if there are any plans showing underground, or above ground, services. These give critical information to guide where to plant new trees – do not plant on top of services that will need to be repaired,

or underneath cables that will get caught up in the growing tree crowns.

Pollution

Pollution in urban areas is indicated by foliage damage to trees and other vegetation, or poor growth of vegetation. Usually, additional signs and information are needed to be able to identify pollution as a significant factor on the management of a site. Additional information may be provided by records, local knowledge, or other surveys carried out in the area.

NATURAL POLLUTION

An interesting example of natural pollution is the build up of bird droppings (guano) beneath large colonies of sea birds that can prevent almost any plant growth.

Biological Survey

Soils

Soil surveys can involve taking samples for analysing back in a laboratory, or using local knowledge to categorize the soil on a site.

Soil texture is a simple characteristic to check and refers to the 'feel' of moist soil as it is handled. The different particles within the soil give it this texture; they are silt, clay or sand. Sand particles are bigger than silt particles, which are bigger than clay particles. The feel of the soil changes as these three constituents are combined in different proportions. Soils with a large proportion of clay or silt particles feel smoother than soils with more sand grains.

Soil texture is determined by rubbing the moist soil between the fingers and thumb and can be determined in the field. It is a subjective technique, but, with practice it can be mastered and, being a quick and simple technique, it has advantages over laboratory tests, which take time and money.

Soil pH is a valuable guide to soil quality. Measuring the pH of a soil is easy with a soil-testing kit and will take only a few minutes. The readings should be taken from a representative sample of spots within the site, as pH can change dramatically over a short distance.

The degree of compaction of a soil is an important factor. Bulk density can be measured by taking standard volume soil samples, drying them out and comparing the dry weight with the sampled volume. A rule of thumb is that compacted soil will resist the force of a knife pushed into it, while the same knife can slide easily into an uncompacted soil. This technique is limited by access to the soil and the knife blade length.

Samples of the soils should be taken from a range of locations within a site, enough to give a clear understanding of the soil variations within it. It is also worthwhile to collect information about the depth of topsoil present. Some soils are very shallow and this limits the volume available to trees. Wherever possible trees should be provided with a minimum of 50cm (20in) depth of soil into which their roots can grow.

Vegetation

To understand a site it is important to look at the existing vegetation, ranging from trees to lichens. The plants present will give important clues to the soil type, soil pH, hydrology and drainage pattern, air pollution and the climate.

Trees are obviously the largest vegetation features and a record of species present and the size of trees is valuable when planning the future of a site. Large groups of trees or plant communities may be able to be classified according to the National Vegetation Classification. However, in urban areas very large tree groups or plant communities are unlikely to be encountered and many species present may be exotic ones brought in for horticultural purposes, or that have 'escaped' from homes or ports and can only survive in towns where the climate is milder than in the countryside. Changes of vegetation should always be noted. These often show where soil variations occur within a site or where soil conditions have been altered to encourage particular plant species.

The patterns of growth on existing trees are important to note. Branch growth and leaf size, colour and density give information on recent conditions and the ability of the trees to cope. Accumulation of dead branches may be a sign of declining growth, especially when found on relatively young trees. Die-back of branches on one side of the crown is often a sign of damage to one part of the root system or trunk. The previous pruning cuts can also be an indication of problems on a particular side of a tree. Although suckering from the base of a tree appears to suggest strong growth, it may in fact be an indication of stress.

Noting the amount of growth achieved by trees over the previous few years is a useful way of assessing site conditions. A small number of typical stems should be chosen and measurements taken from the tip to the previous year's terminal bud, giving a reading for the tree's annual growth. It is often possible to see clearly the incremental growth on a branch for the past five to ten years and, even without measurements, periods of good and poor growth can be compared (see Trowbridge and Bassuk, 2004).

Fauna

Animal and insect pests can have a dramatic effect on trees and any clues about their presence are valuable for site assessment. Signs of animal damage may range from humans breaking semi-mature trees to foliage and shoot browsing by deer or rodents. Insects will also affect foliage but their presence can be deduced by spotting patterns of die-back and regrowth on larger trees.

Existing Management of Flora and Fauna

On larger areas, information about previous or existing management is useful to collect. Old agricultural patterns can give clues about soil conditions, for instance, and this may be a feature that is appropriate to bring forward in some way into the future management of the site in order to retain a link with the past. The most likely feature to be encountered in

A pin oak showing signs of iron deficiency or high pH.

this category is a hedgerow or field boundary. The cultural, historical and habitat value of these is being recognized more and more as they become scarcer, so any survey should record their presence, position and condition wherever possible.

It may also be that a site includes a view over the surrounding area and that trees beneath it are routinely cut back to limit encroachment.

Human Survey

Site assessment must consider the human perspective and look at existing land uses, but should also consider any traditional uses and cultural significance that could have a bearing on future management of the site. Access and internal roads and paths should be included, as these will undoubtedly affect how the area is used in future. Public perception of the site is a vital ingredient and site users, neighbours and any other stakeholders should be consulted in some way when attempting to gain an understanding of how people feel about the area and the different ways in which the site is used.

Visual Survey

Assessment intended to arrive at the best use and management of a site or area will have to consider the effect of the site beyond its boundaries and also views within, into and

Shoots showing clear annual shoot extension.

out of the site. Internal landmarks should be identified and incorporated into plans and the visibility of these features outside of the site should also be considered. The visual quality and character of the site and surroundings are crucial elements to consider when deciding how to manage a site.

SITE MODIFICATION

Once a site has been evaluated, a range of management options are likely to be considered. The basic choice is either to work with the existing site and soil conditions, using appropriate plants that can cope, or to choose preferred plants and modify the conditions to suit them. Site modification often involves regular, periodic inputs to maintain conditions that allow preferred species to thrive. Often in towns the soil conditions are so poor that some form of remediation is essential to provide conditions that favour any plants.

In urban areas, typical situations that require alteration are those where waterlogging is frequent; where soil compaction is known or suspected; or where poor materials make up a significant part of the soil, maybe due to fly-tipping or toxicity. Soil acidity or alkalinity may be so limiting that action is needed; landfill sites may suffer from methane escape; and soils may be very weedy or uneven. Often, in very built-up areas the main issue will be the provision of a sufficient volume of rootable soil for tree growth and provision of drainage and irrigation measures to ensure that trees can thrive.

On large sites that become available for tree planting, it is likely that a range of activities will be proposed for future use. Here, attention should be given to the form of the land prior to any other operations so that the way water will move over and through the site is understood. Slopes help to shed surface water and direct it to areas where it can be used. Gentle slopes avoid problems with erosion, while short slopes prevent surface water building up momentum and causing problems and erosion lower down.

SPECIES CHOICE

The choice of species when considering tree planting is obviously dependent on what information has been gathered during the various surveys. Soil conditions and climate are primary constraints on which trees will thrive in an area, but there are further limitations that need to be considered.

Urban conditions are generally a challenge for trees, so more hardy species with some tolerance to pollution, moisture stress and disturbance will be favoured. Toughness may also be a consideration where vandalism is common. The soil volume available during a tree's growth is critical, as any tree restricted to exploiting say a mere $5m^3$ of soil will not be able to grow above a quite limited size. So, if a tree is to be planted into a pot above ground level, or into an island of soil surrounded by concrete, it will have a very limited lifespan and part of the plan must be to remove and replace it fairly frequently.

An understanding of the history of a site and the benefits provided by existing trees should influence any new planting scheme. Trees that are valued by a community should be chosen before those that are seen as having major disbenefits. For example, trees close to buildings casting dense shade may be objected to more than those casting only dappled shade. Similarly, trees with heavy seeds that could cause injury, or fruit that stains easily and makes footpaths slippery should be considered from the community's perspective.

The objectives of any tree strategy come into play at some point. Small cherry species, for instance, will not be able to provide the benefits of scale, shade or carbon fixing that large-growing species can. The latest knowledge about climate change must influence species choice. Where conditions are expected to favour some species and put others under stress decades from the present, it is wise to act prudently. One consequence of climate change that we are already dealing with is an increase in pests and diseases. Reliance on a very limited range of tree species could result in an exposed and vulnerable tree stock in future.

So, species choice is a process of identifying those trees that are likely to have the capability to thrive where they will be planted and that will be acceptable to the local community, providing easily recognized benefits.

REFERENCES

Bradshaw A., Hunt B. and Walmsley T., *Trees in the Urban Landscape*, (E. & F.N. Spon, 1995).

Britt C. and Johnston M., *Trees in Towns II: A New Survey of Urban Trees in England and Their Condition and Management* (Department for Communities and Local Government, 2008).

Kozlowski T.T., Kramer P.J. and Pallardy S.G., *The Physiological Ecology of Woody Plants*, (Academic Press, 1991).

Meteorological Office, Historic UK Climate Records (Meteorological Office, 2010).

Miller, *Urban Forestry* (Prentice Hall, 1998).

Schroeder H., Flannigan J. and Coles R., *Residents' Attitudes Toward Street Trees in the UK and US Communities*, Arboriculture and Urban Forestry (ISA, 2006).

Trowbridge P. and Bassuk N., *Trees in the Urban Landscape: Site Assessment, Design, and Installation* (Wiley, 2004).

Van Wassanaer, P., *Sustainable Urban Forest Management Planning Using Criteria and Indicators* (Seminar at 44th Arboricultural Association Annual Conference, 2010).

At the edge of towns development encroaches on agricultural land and trees.

chapter five

Planting, Establishment, Pruning and Maintenance

PLANTING AND ESTABLISHMENT

Planting is the action of bringing a tree to a new location and setting it securely into the soil in order for it to grow into a mature specimen. Establishment is the process of enabling the tree to become accustomed to its new situation and able to access sufficient resources to become self-sufficient for growth and stability.

Planting and establishment go together. One without the other is only half the job. Planting without establishment gets the trees into position, but does not deal with the other factors that can make or break a planting scheme. If the end objective is to grow a thriving tree, planting by itself will not get the job done. You have to go back and carry out a few, simple operations to ensure that the tree is settled and growing well. My guess is that the majority of failed tree-planting programmes over the last thirty years have been the result of a failure to carry out establishment operations in the two years following planting. It is as important as that; without establishment, half your planting effort (and maybe more) is likely to be wasted.

A new British Standard is being developed that aims to draw together all the experience of the nursery, landscape, horticultural and arboriculture industries to set out practical standards for tree production from the nursery to the time when trees are established and providing benefits to their surroundings.

Nurseries and Trees

The process starts in the tree nursery. The nurseryman uses his art and skill to raise robust plants of good quality that have the capability of growing well. These plants need to be selected for use (and poor-quality plants rejected), then transported to either a holding area or the planting site. This move in itself provides ample opportunity for the nurseryman's work to be undone by poor handling, poor protection from wind and sun and lack of care.

Tree Sizes

Trees are grown either from seed or cuttings and are cared for in a nursery until they are ready for use. The nurseryman controls the growth and spread of roots by either transplanting the trees periodically, or undercutting them using a wire that passes beneath and between the soil beds in which the trees are growing. The growth of the stem and shoots is controlled mainly by pruning, but chemicals can also be used.

Trees spend a variable amount of time in a nursery, or nurseries, dependent upon the specification of the final product. Transplants

Bare root whips. (Photo courtesy of Hillier Nurseries)

and whips are small trees that have been growing only one to three years and have been transplanted at least once. They are used mostly in forestry or large-scale plantings. Whips are transplants of up to 1.25m (4ft) in height and without significant side branching of the stem. They are small, light and easily planted. Transplants are usually sold as bare-rooted, but are usually supplied in co-extruded polythene bags, black on the inside and white on the outside. Whips are usually sold with their roots only in this kind of bag. Small quantities of whips may be sold in containers.

Feathered trees are defined in the National Plant Specification as trees usually with an upright central leading shoot with evenly spaced and balanced branches down to near ground level, according to species. They are up to 2.5m (8ft) tall and usually sold in containers or root balls. Maidens are trees that have had a bud or section of stem grafted onto the base of a rootstock plant and allowed to grow for a year.

Standard trees are those that have a clear, upright stem and a crown of lateral branches that form a shape natural for the species. Standard trees range from light to extra heavy, depending upon height and girth, and are the usual type of trees specified on amenity planting schemes in urban areas. At the large end of this range they may be referred to as advanced nursery stock. They are big enough to be immediately noticed by passers-by, but small

A co-extruded bag for bare-root standard tree. (Photo courtesy of Hillier Nurseries)

A standard tree nursery line. (Barcham Trees nursery near Ely)

Scots pines, 14–16cm (5.5–6in) in 200ltr (44gal) containers. (Barcham Trees nursery near Ely)

enough to be transported and planted easily. Standard trees should have been transplanted within the nursery between two to four times before they arrive at the planting site.

Semi-mature trees are defined as having a height in excess of 5m (16.4ft) and/or a stem girth measurement of 20cm (8in) or greater. All girth measurements on nursery trees are taken at 1m (3.3ft) above the growing medium. These trees are regularly transplanted in the nursery. The National Plant Specification states that transplantation should have been carried out four times for trees at the lower end of the range and six times for bigger specimens.

Standard and semi-mature trees are sold bare-root, or in containers or root balls.

Because of the weight and size of these trees, they require specialist equipment for transport and planting. Root-balled plants are grown in the nursery ground and wrapped in hessian or polythene when they are lifted. Containerized trees are those that have been lifted and put into a container, or transferred into a larger container, but have not yet made substantial new root growth. These trees are not ready for sale until that time, when they are called 'container-grown'.

There are a few categories for more specialist trees, such as multi-stemmed and bushy trees where a less formal effect is required, and also it is possible to order trees that have been 'tailored' for a specific purpose. For instance

Lime, 30–35cm (12–14in) girth in 500ltr (110gal) container. (Barcham Trees nursery near Ely)

Lifting a semi-mature tree. (Photo courtesy of Hillier Nurseries)

A semi-mature tree prepared for transport. (Photo courtesy of Hillier Nurseries)

espalier trees have a balanced branch structure extending horizontally on opposite sides of the stem in a vertical plane. Fruit trees are often grown in this way. Pleached trees have a similar branch structure to espaliers, but these branches are much higher above ground in order that a raised-hedge effect can be produced. Pollard plants have been cut back to a height of around 2m (6.5ft) (or to another

TREES NOT TO BE PLANTED AS BARE-ROOT PLANTS – ALWAYS SPECIFY THESE TREES TO BE SUPPLIED AS ROOT-BALLED OR CONTAINER-GROWN (COURTESY OF HILLIER NURSERIES)

Acer capillipes	Snake-bark maple
Acer davidii	Snake-bark maple
Acer griseum	Paper bark maple
Aesculus indica	Indian horse chestnut
Amelanchier	June berry
Betula	Birch
Catalpa	Indian bean tree
Cercidiphyllum	Katsura tree
Fagus	Beech
Ginkgo	Maidenhair tree
Liquidambar	Sweet gum
Liriodendron	Tuliptree
Parrotia	Persian ironwood
Quercus	Oak
Robinia	False acacia

specified height). Parasol trees are pruned by cutting back the central leader and training the branches to grow over a shaped frame in the nursery to form an umbrella-shaped crown.

Conifers are classified differently from broad-leaved trees. Most conifers are used either for forestry planting, or as garden ornamental plants. Broad-leaves are defined by their size and shape; conifers are defined by their height alone. The National Plant Specification uses other criteria for checking the quality of supplied plants, but, in general when ordering conifer trees, you just need to state the required tree height.

Trees are vulnerable to drying out, mechanical damage and frost between being lifted from a nursery bed and planting in their final position. They must be stored in a cool, shaded place for as short a time as possible and then transported as quickly as possible in covered vehicles to the planting site. There

they should be planted as quickly and carefully as possible. If the trees cannot be planted immediately, they must either be stored in a cool, shaded place temporarily, or, for bare-rooted trees, 'heeled-in', that is, planted for a short time in a prepared trench. This keeps their roots covered and protected from frost and wind.

Because of the effects of poor handling and transporting on subsequent tree growth, it is important for any tree planter to check the quality of trees received. Poor-quality, frosted, dried out or poorly formed trees will always underperform against good-quality nursery stock. Nursery work and the production of trees are often separated from planting and maintenance, which breaks the natural involvement and commitment of nurserymen who want to see their trees thriving in their planted positions.

Barcham Trees in Cambridgeshire are currently developing trees with greater taper and sturdier characteristics than has been common up to now. Lack of taper in nursery trees can result from specifications for tall, clear-stemmed trees for large planting schemes and from the natural desire to raise as many trees as possible on a limited area. Reducing tree densities in the nursery can help to increase stem taper and robustness in trees.

Planting

Once on site, the trees need to be planted efficiently and professionally. A clear plan should be prepared in advance indicating planting positions and all tools and materials should be in place so that the trees can be planted quickly.

Planting is a traumatic change for a tree. It is taken from a controlled nursery environment and put into a new setting with different shelter, light, water supply, soil type and exposure to damaging pollution, and all this with only a fraction of its root system in the case of bare-root and root-balled stock, or with a small, vulnerable root ball even if the tree is container-grown. No wonder trees can suffer 'transplant shock'.

Planting specifications are included in British Standards BS 4043 and BS 4428. The planting hole needs to be a minimum of 1.5 times the root-ball size, container or bare roots of the tree. The roots should be firmly pressed into contact with the soil. Trees should not be planted into waterlogged soil or frozen ground. In general, the tree will be at the mercy of the surrounding soil type, with its characteristics of structure and texture, its drainage and slope. However, on some sites there may be no existing soil and so imported soil will be needed.

Peat, or a peat substitute, can be useful in softening the soil around tree roots. Many soils are stony and can abrade roots if firmed around them too much. Because peat, or peat-like material, is high in organic matter, it has good water-holding characteristics, which is another benefit.

Once planted, trees need to be watered to ensure the roots do not dry out. At planting time in the dormant season enough water should be added to make sure the soil is moist through to the bottom of the Rootball. If planting is done using containers at the start of spring or further into the growing season watering will need to be carried out regularly to keep the tree healthy.

Where there is impeded drainage this should be remedied before planting and waterlogged soils should be avoided and not planted.

Take time to water each tree and avoid pouring on water that runs off and away from the roots. Forming a soil ring around the edge of the rootball creates a basin to hold water.

After watering, about 50mm of mulch should be spread to cover an area at three times the radius of the rootball.

After-Planting Care

A well-planted, good-quality tree will probably require attention for two to four years before it becomes self-supporting. During this time the operations that need to be carried out are:

- watering
- weeding/mulching
- checking guards, stakes and ties
- checking the need for fertilizing.

Watering

During the first few growing seasons a young tree is likely to need help in meeting its water needs. This need increases as the soil volume available to the tree decreases; at planting time, the roots only occupy a very small volume. A tree in a very restricted planting space (possibly in a street or shopping centre) will be more vulnerable to drought than a similar tree planted in an open area (without grass competition) with a larger soil volume surrounding it.

Regular watering is clearly the ideal situation, with adequate water supplied each time to reach field capacity and frequency of watering being sufficient to avoid the soil drying out severely. But achieving this can be tricky where money, expertise, time or suitable contractors are in short supply. If a watering regime needs to be specified, typically for a contract document, it is important to factor in the likely water deficit experienced in the area with the conditions around the trees. Ultimately, what a contractor (or any watering person) needs to know is how often to water and how much to supply. One general rule is to provide 1ltr (1.75pt) of water per day for each square metre of leaf surface. So, weekly visits could be provided with 7ltr (12pt) of water per square metre (11sq ft) on each occasion. If watering were to be done once every two weeks the amount would be 14ltr/m^2 (25pt) each time. Most newly planted standard trees have a leaf area of between 1m^2 and 3m^2 (11–33sq ft). Therefore, it can be assumed that the rate of water loss of most recently planted trees of this size will lie somewhere in the range of 1–3ltr (1.75–5pt) per day (see Bradshaw et al., 1995). Watering should be provided throughout the whole growing season, but it is particularly crucial during spring and early summer, when most growth happens.

When planting semi-mature trees, or planning tree planting in very urban environments, information about the volume and characteristics of the soil is crucial. Trowbridge and Bassuk (2004) detail a calculation that firstly considers the ultimate 'design' size of a tree, then determines what volume of soil that tree will need, taking into account its water needs and the water-holding capacity of the soil. These last two parameters can then be used to specify a watering regime, independent of natural rainfall (see Chapter 7). To factor in the natural rainfall, the expected length of dry periods (with no rain) must be known. In the UK this information is available from the Meteorological Office, covering the whole of the country in 5km grids. The information details the consecutive number of days with rainfall below 0.2mm and number of days with rainfall at or above 1mm along with the total rainfall on those raindays.

The more a local climate is understood, the more precise can be the water provision. For instance, in an area that often experiences droughts in spring the watering regime should be adjusted to meet this need. Where a water deficit is usual only at the end of the summer watering may be less critical, but it must always be borne in mind that newly planted trees are more vulnerable to drought than well-established, older trees. Information about the climate of a particular area is provided in the *Soil Survey of England and Wales Bulletins* and in *The Agricultural Climate of England and Wales*.

Watering beyond the initial establishment phase is not usually effective; once a tree root system has grown outward from the tree for a few years it is very difficult to decide where to water to favour the roots. There are times when larger trees may need to be invigorated by watering, but these will depend upon an assessment of all the circumstances.

Weeding/Mulching

Mulch provides a buffer between the air and the soil; it reduces water evaporation and helps to maintain the current temperature of the soil. It also shuts out light to the soil surface, thereby impeding germination.

The purpose of weeding is to prevent other vegetation from robbing a newly planted tree of resources such as water and nutrients. Where annual weeds grow tall they can also smother small trees by the end of a growing season. There are several ways to combat this threat to young trees:

- chemical weeding
- hand weeding
- natural mulches
- synthetic mulches
- inorganic granular mulches.

Weeding can be done by hand, pulling up all weeds around a tree, or by using tools such as sickles, slashers or hoes. This is time-consuming and repetitive work, but it avoids adding chemicals to the soil around the tree. If the only method of weed control is hand weeding, it is likely that it will need to be repeated several times during the growing season.

Chemical weeding is usually more cost-effective because it is necessary only once each year. Chemicals used as herbicides need to be carefully chosen to target the main competitors to young trees and to be harmless to those trees. Many herbicides are licensed to be used only in specific situations (for example, away from watercourses) and so the full specifications of any herbicide must be studied before use.

A mulch blankets the soil and, by doing so, limits the growth of weeds. Weeds tend not to germinate in very low light conditions, so mulch can help to prevent them from growing to compete with young trees. Organic mulch usually consists of bark or wood chips. Mulch limits water loss and temperature change, therefore the best time to apply most mulches is in the autumn when the soil is warm and the water content is high. Natural mulches break down over time, or are blown away or tossed around by birds, so they must be reapplied

A mulched tree. Mulch must be added annually; this tree is ready for fresh mulch.

annually. It also helps to plant the trees in a planting pit that is slightly below the level of the surrounding soil, so that the mulch is naturally held in this depressed 'saucer'.

Usually, 50mm (2in) of mulch is a sufficient depth. Up to 75mm (3in) is acceptable, but greater depths can have adverse effects on young trees. It is important not to stack the mulch against the base of the tree, as this can smother the bark and lead to problems. When applying mulch, always ensure that the depth decreases to nearly zero around the base of the tree's stem.

An alternative to natural mulch is one that covers the ground using a synthetic material, for example plastic sheeting or old carpet, or, in imitation of natural mulch, rubber-tyre chips. The principles are the same, but the cover only needs to be reapplied if it is damaged or breaks down over time (such as plastic in UV light). Plastic covers can become a problem if left around a tree for too long due to the difficulty of water passing through them. Even though the water loss beneath them is reduced, it will eventually lead to drought as the tree draws up water for growth.

Inorganic mulches include pea shingle, gravel and stone chips. These can do the same job as natural, organic mulches and do not break down in any timescale that we can observe. They may need replenishment over time due to sloping land or frequent pedestrian traffic. They should be used in the same way as organic mulches, limiting depth especially around the base.

Guards, Stakes and Ties

Guards around the base and stem of newly planted trees are sometimes necessary to prevent damage from animals. Young trees can attract unwelcome attention, be that from bored youths, household pets or wild mammals. Vandalism by people or pets is usually very site-specific and needs careful thought, creativity and persistence to overcome, while rodents can be deterred from snacking on small trees by the

use of rabbit guards. These flexible tubes keep rabbits and voles away from the vulnerable bark of young trees. They only need to last a couple of years on most sites.

Stakes are temporary supports for young trees planted into new sites. Their purpose is to hold the root system firm enough to allow root hairs to grow out from a root ball into the surrounding ground and to become woody roots. If a tree rocks in the planting pit, the root hairs cannot anchor themselves beyond the root ball, with the result that the tree remains unstable.

Stakes and ties, if applied properly, are useful for two or three years. After this time, most trees will be well-enough adapted to their new home and their roots will have taken a firm-enough grip on the surrounding soil to allow their removal. Some trees may take a little longer (*Sorbus*, for example establishes slowly), but, in general, if a tree still needs its stake and tie after this time it was probably not planted properly – consider digging it up and starting again.

Sometimes, with larger trees, some support for the trunk is needed but this is a secondary consideration. Simple wooden stakes and rubber ties are relatively cheap. For semi-mature trees ground anchors are usually necessary, or else guys attached to the trunk. Sometimes both systems are needed.

Whatever type of staking system is used, it must be removed before it starts to damage the tree, threatening its future. Stakes and ties left around a tree will affect the incremental growth of the trunk and become tourniquets blocking water and starch transport around the tree and affecting the strength of the trunk. If you are not committed to returning to a tree to remove the stake and tie, it is best not to use them at all. If that is the case, use the money not spent on this hardware to buy and plant more trees.

Underground guying systems aren't removed. Only the above ground parts are dismantled when they are no longer needed.

Checking the Need for Fertilizing

At planting, only a small amount of slow-release fertilizer should be added to the soil surrounding the roots. This is because the root system must grow and increase the tree's ability to absorb more water before it can benefit from fertilizer. In most natural soils in the UK there are sufficient nutrients for tree growth, certainly for the establishment stage. However, in urban areas a natural soil is not always present. Where growth of young trees appears stunted or discoloured the reasons should be investigated. It may be that the symptoms give a clear idea of what is lacking in the soil, or a soil analysis may be needed. Although soil analysis is an added expense it is often worth doing, preferably before planting, to ensure that the planted trees are given what they need to grow and so provide a return on the investment of planting. If any new trees need to be planted as replacements, it will also make sure they don't suffer the same fate. Nutrient deficiencies can take a year or more to show up in young trees, therefore it is usually best to wait a while before adding fertilizer.

Checking the pH of a soil can be done quickly and inexpensively with a pH testing kit. This is important because of the effect of pH on nutrient availability.It is possible to test the capability for growth of the soil using a Solvita kit, which measures the degree of microbial activity.

To allow the tree access to soil resources, competing plants should be kept away. Mown grass is a particularly aggressive competitor for resources. No grass should be tolerated around the root ball or root zone of a newly planted tree. Other weeds may not be as competitive as grass, but their removal will usually benefit the new tree roots. Planting trees close to existing trees can also be counter-productive, as the older trees' established roots will get to the water and nutrients first. A few trees and shrubs exude chemicals through their roots to inhibit the growth of competing plants and this can affect newly planted trees.

Where new soil needs to be imported it should be specified to conform to BS 3884:1994

An underground guying system used to secure a river birch at Oxford Botanic Garden.

Topsoil. This British Standard has three grades of topsoil – premium, general purpose and economy. In most cases, the general-purpose grade will be sufficient for tree growth, although in highly constrained urban sites it is best to buy the best-quality soil possible. Quality control is always needed to prevent contaminated or weedy soil being brought onto a site. This can cause major problems, resulting in trees struggling to get established or even dying, and can lead to added costs in removing contaminants or aggressive weed plants.

Where trees fail to thrive on a site, the problem is likely to arise from some breakdown in quality in the planting and establishment process. In some cases, it may be possible to identify and remedy the situation, but often if the trees are not thriving after three to four years the best option will be to remove them and start again.

PRUNING

Pruning is the planned and controlled cutting of branches, stems and roots for a specific reason. That reason may be to control growth, flowering or fruiting and, where carried out, pruning can have a major influence on the subsequent form of the tree. Pruning should enhance the health and vitality of a tree and increase the contribution it makes to the landscape. The process begins in the nursery.

In the nursery, the reason for pruning is to maintain a good root-to-shoot ratio in young plants; in newly planted trees formative pruning is carried out to encourage a healthy, well-spaced branch structure that avoids major defects that would require significant attention in future. In established, larger trees, pruning is carried out to remove hazards to people and property and to regulate the size of the tree in order to avoid problems from encroachment into spaces needed for other purposes. Pruning of orchard trees is carried out to encourage fruit growth.

With the right tree planted, pruning can be kept to a minimum. A tree well suited to the ground and environmental conditions will develop fewer problems that require pruning and a tree in scale with its surroundings is less likely to need regular pruning to contain its size. But existing trees are not always of the best species, new pest and disease challenges arise and buildings and structures are installed closer to trees that once had much more space. For these reasons, pruning will always be an important tool for urban tree management.

Pruning Young Trees

In the nursery the emphasis is on producing a good-quality plant with a strong, healthy leading shoot and well-spaced lateral branches. Only where an open-crowned tree is the objective will the leading shoot be routinely removed or pruned.

Where defects are found in young trees they should be addressed as soon as possible. Leaving defects may save a few pennies initially, but it can cost substantial amounts if the tree reaches a large size with a poor branch structure. This can ruin the shape of the tree, make it vulnerable to storm damage and undermine its ability to meet the objectives for which it was planted.

Pruning at the time of planting is ideally carried out by the nurseryman before delivery to the planter. Cutting back of side branches by skilled nursery staff avoids the risk of poor pruning by contractors on site, makes the transport operation less likely to damage the trees and reduces destabilizing and desiccation by the wind. In addition, the careful reduction in number and length of branches reduces the stresses on the tree during the first growing season. Pruning at this time invigorates shoot growth and reduces the size of leaves.

There has always been some debate about how and when to prune trees at planting time. There is an argument to say that the tree should be allowed to prune itself and that artificially cutting it back adds to the transplant shock. However, not pruning will lead to the need within one or two years to prune out the dead, disfiguring twigs.

Whatever pruning is carried out, the purpose is clear – to bring the tree into a balance between its roots and crown as quickly as possible. The root system is severely reduced in bare root or root-balled trees and this affects water supply to the branches. A healthy branch structure is needed to power new growth of shoots and roots, so pruning too much can affect a tree's ability to grow out of transplant shock. The research reported in Bradshaw *et al.* (1995) found that pruning of newly planted trees was very damaging if carried out during the first growing season once the leaves had emerged.

Formative pruning, commencing a few years after planting, continues the nursery work and is minimal where the nursery was successful in removing crossing branches, tight forks and competing leading shoots. A period of rapid growth following establishment is often very noticeable where good-quality trees settle well into their new surroundings, especially with faster-growing subjects such as silver birch (*Betula pendula*). A completely new head is often formed, which is good, provided that it has developed from the original leading shoot. The lower branches of the old head may be gradually removed, which is best done in late summer, over three or four years if necessary.

Branches of broad-leaved trees should be well spaced on the trunk, both vertically and around the circumference, as they are then more likely to have strong attachments than when several branches are growing close together.

Pruning Semi-Mature Trees

Semi-mature trees that are very slow growing may begin to accumulate deadwood in the crown. Pruning can be used to stimulate new growth, but care must be taken to avoid damaging an already struggling tree and pruning must include the careful selection of good growing points from which new growth can spring. This invigoration is best done over an extended period with, first, a summer pruning to remove dead and weak growth when these are easily recognized. Once a tree has responded to this treatment there should be more new growth and more options to enable the pruner to select strong shoots to take over from stunted branches.

Pruning can be carried out below ground, or at ground level, as well as in the tree crown. Girdling roots can form around the whole of the base of a tree or around only part of the circumference. Such roots can restrict the passage of water, nutrients and sugars up and down the stem and may produce a structural weakness in the trunk. At the semi-mature stage it may be practical to cut out a defect such as this. When a tree is mature the damage caused by this work is likely to outweigh any advantage. Suckers are shoots arising from near or below ground level. They are generally scruffy, get in the way easily and drain resources away from the crown of the tree. If they form on a young or semi-mature tree they should be pruned back to their point of origin.

Pruning Mature Trees

On mature trees pruning becomes a more complex, costly and dangerous operation that requires skilled attention from tree surgeons working at heights with chainsaws, cranes and other specialist equipment. In general, such pruning is likely to be needed for one or more of the following reasons:

- to raise the crown above ground to enable passage of people and vehicles beneath or to improve light and air access to ground level
- to prevent trees from growing beyond the space allocated to them
- to reduce the degree of shading from the crown
- to remove diseased or hazardous branches
- to shorten branches that are or may become hazardous
- to restore a desired shape to the crown after damage from disease, pests, climatic events or accidents.

The operations carried out on these trees fall into the following categories and are explained in BS 3998:2010 *Recommendations for Tree Work*:

- **Removal of deadwood** Decaying wood is a valuable ecological resource and should only be removed where it poses a definite risk to people or property. Other unhelpful materials in a tree, such as wires, string, timber and smothering plants, should also be removed if necessary for human safety and tree health.

- **Crown thinning** This is the selective removal of branches to increase light penetration and air movement. It should result in an even density of foliage throughout the crown. Trees treated in this way often need further work periodically.
- **Crown lifting** This involves the removal of low-growing branches to provide clearance beneath the tree, usually for vehicles or pedestrians. It also improves light penetration and air movement beneath the tree crown. The work ideally prunes secondary branches rather than primary ones, as this avoids producing large-diameter wounds. A maximum of 15 per cent of the live crown height should be removed and the pruned crown should still make up two-thirds of the tree height.
- **Crown reduction** This is usually achieved by shortening the leading, most vertical stems and shaping the surrounding, lateral branches to approximate a natural-looking crown. This work is necessary where a tree outgrows the space allocated to it, or where damage to the upper crown requires cutting back of one or more of the main stems.
- **Branch pruning** Such pruning is carried out to specific limbs for specific reasons, usually due to damage or disease, or possibly due to the branch extending too close to nearby structures.
- **Pollarding** This is a pruning operation that shortens the branches of a tree back to previous pruning cuts, where a 'knuckle' or 'knob' has formed. In this way, the whole of the crown is removed and the operation is carried out periodically to maintain the tree at a chosen size, or for a chosen architectural purpose. Pollarding begins when trees are of small size and when pruning cuts are of small diameter. Cutting back all the branches of a mature tree that has not been 'trained' as a pollard introduces far larger wounds into the main trunk and is called 'topping'. Topping should be avoided if at all possible.

Specialized pruning techniques have been devised to produce dramatic effects: screen trees are pruned to form a hedge effect at a set height above ground level, supported on the trunks of the trees; the branching of espalier trees is directed into only two dimensions, reaching out from the stem as horizontal laterals or in a fan shape; pleached trees (meaning interwoven) combine the two-dimensional effect of espaliers with the raised crown of the screen trees; and topiary is a form of repeated cutting back to a set point to create unnatural shapes from the tree crown. Cloud trees are formed by pruning branches so that groups of shoots are encouraged at branch tips and these are kept compact by regular trimming. This produces a tree with a number of foliage clouds, each in its own space along the branches.

Pruning Veteran Trees

Veteran trees are trees that have become important locally, regionally or even internationally, due to their biological interest, aesthetic or cultural value and age. They are trees nearing the end of their natural lifespan, or that are old relative to others of the same species. The age and size of these trees makes working on them a delicate business. Old trees are generally intolerant of disturbance or changes in their surroundings and extra consideration must be given in urban areas where such trees remain and are planned to be conserved for as long as possible.

The general rules and principles of pruning are still valid, but, because all the live parts of the tree are likely to be significant for its well-being and much of the deadwood is likely to be important for ecological diversity, care is needed to minimize necessary work and to identify threats to tree health and opportunities to improve conditions for the tree.

Two main reasons for pruning veteran trees are: either the work is necessary to safeguard people and property around them (remedial); or it is necessary to bring the tree back into some form of active management that will extend its safe, useful life. Remedial work is a one-time treatment, unless establishing a new, smaller crown, whereas restoring a former management regime, such as pollarding, will require periodic operations for the rest of the life of the tree.

General Considerations

Size is important when considering pruning wounds, and relative size even more so. A rough guide to use is that the removal of a branch should preferably create a wound not larger than about one-quarter to one-third the diameter of the parent stem or branch. Knowledge of the differences between species is needed so that pruning leads to natural-looking trees that can remain healthy and safe on a site. Knowing when not to cut some trees is equally important in order to avoid increasing problems for the tree.

Pruning for health and safety reasons cannot be dictated by time of year and physiology of a tree, but programmed tree works can take into account such details as trees that 'bleed' at the end of the winter (for example, birch and maples) and those that need to be pruned when in full leaf (for example, cherries and walnuts). Veteran trees in particular need to be understood in regard to their species characteristics as well as their age, so that appropriate pruning can be planned.

MAINTENANCE AND CARE

All pruning, of whatever kind, should seek to minimize the damage to the tree. Correct pruning cuts, careful attention to detail and knowledge of the characteristics of different trees are all needed to prevent an erosion of the value of a tree. However, there are other things to consider when managing and caring for an urban tree resource.

Individual members of large groups of trees, woods and copses are unlikely to be pruned unless they are possibly at the edge of a

A veteran sweet chestnut tree, Ashton Gate, Bristol.

road or near a structure. A more appropriate operation is thinning, in which defective trees are removed for the benefit of those remaining. As trees grow they obviously take up more space, which means that fewer can comfortably be accommodated in a set area. Selection is therefore needed to pick out some for retention and some for removal. It is not always the best-formed trees that will remain, although this would be the ideal choice. Rather, the contribution of the tree group to the surroundings should be clearly understood and selection aimed at maximizing this benefit. So, for instance, it may be that a worse tree would be retained because it is in a better position than a well-formed neighbour.

Thinning is an important tool for woodland management and the choice of which trees to remove, how many to remove and frequency of thinning cycle needs careful consideration.

Taking out the wrong trees leads to a build-up of defects that can be costly to fix if the trees are damaged. It can also lead to the trees not being vigorous enough to use the space around them. This can be the problem if too many trees are removed at once. The wood then looks bare and weeds can colonize the area, thereby reducing the visual impact of the tree feature. If the time between thinning operations is too long, the trees can outgrow the space between them and grow into tall, spindly specimens that cannot adjust when their neighbours are removed.

Unplanned fires can be frequent in open areas and the only response to this is to clear up afterwards, taking remedial action if trees are obviously damaged. Fires cause damage by scorching leaves and trunks with the flames and also by the intense heat, which can severely affect all parts of trees, including root systems. The danger of damage from fires near to trees means that there should be no planned fires close to them. A bonfire should be in the middle of a field, not near a hedgerow and fires on development sites near trees should be outlawed.

As trees age, their branches and stems increase in size and weight. Inspections may identify defects such as subsiding branches or weak forks. In some cases, these situations can be managed without pruning. Low branches may be supported by props of timber or metal. Weaknesses in the crown may be supported by cables placed around co-dominant stems, for instance, or possibly connecting a low branch to the main stem. Wherever these options are appropriate, the safety of people and property must be taken into account and follow-up inspections carried our regularly to monitor the effectiveness of the work.

It is still quite common to find trees in parks that are filled with cement or even bricks. This was a recognized method of treating hollow trees in the early twentieth century. But it is no longer recommended as it does not prevent decay and the hard material inside the hollow abrades and scratches the inner surface of the wood as the tree flexes. Hollows should be left open to aid inspection, to encourage wildlife and to allow the tree to produce reaction growth.

Where water pockets are found they are best left intact if the tree is to be retained (following careful inspection), as the development of pathogenic fungi is inhibited where moisture content is high and oxygen is lacking. Similarly, the old practice of using wound-sealing paint has been found to have no significant benefit in preventing fungal colonization or encouraging callus growth. The main exception to this is for protecting cherry trees from leaf diseases, when a mercury-based sealant can be beneficial.

REFERENCES

Bradshaw A.D., Hunt B., Walmsley T., *Trees in the Urban Landscape* (E. & F.N. Spon, 1995).

BS 4428, *Code of Practice for General Landscape Operations (Excluding Hard Surfaces)* (BSI, 0989).

BS 4043, http://shop.bsigroup.com/ProductDetail/?pid=000000000000201270 *Recommendations for Transplanting Root-Balled Trees* (BSI, 1989).

BS 3998, *Tree Work – Recommendations* (BSI, 2010).

Findlay D.C., Colborne G.J.N., Cope D.W., Harrod T.R., Hogan D.V., Staines S.J., *Soil Survey of England and Wales, Bulletins* (SSEW, 1984).

Landscape Industry, *National Plant Specification*, Ministry of Agriculture Fisheries and Food, *The Agricultural Climate of England and Wales* (HMSO, 1984).

Read H., *Veteran Tree Management* (English Nature, 2000).

Trowbridge P. and Bassuk N., *Trees in the Urban Landscape: Site Assessment, Design, and Installation* (Wiley, 2004).

Watson G.W., Himelick E.B., *Planting Trees and Shrubs* (ISA, 1997).

chapter six

Monitoring and Assessment

It is human nature to wait until a job has to be done before it is tackled, then to allocate just enough resources to it to satisfy the person who gave us the job in the first place. The same principle is readily seen with tree management. Many tree populations are crisis-managed, with only trees that are blatant hazards or get in the way of other activities being dealt with.

Bartlett Tree Research Laboratories has found that crisis tree management accomplishes work on only about one-third of trees that would be dealt with in a planned programme. Clearly, it is more effective to plan a tree-management programme than to trust to crisis management and such an approach will inevitably include tree inspections.

Large plane tree in a busy urban location. Such trees should be monitored regularly.

WHY TREE INSPECTIONS ARE NEEDED

Effective management of trees requires that they be inspected at intervals; ideally regular, programmed intervals. Inspections provide information on which plans and strategies can be based and subsequent inspections throughout the lifespan of trees can check if they are meeting the objectives set for them.

When trees are young and small, inspections will be mainly to check on growth. But as trees increase in size, the safety of people and property becomes a primary objective of inspections. Dangerous branches and unstable trees should not be tolerated in busy urban areas. There is no benefit to keeping a safe tree if it is not a positive feature; equally a large, well-loved tree is still an unacceptable liability if it is about to collapse and sited where it could do damage.

Trees are amazingly resilient and there are usually warning signs before they fail, break up or collapse. Most structural problems in trees can be identified through careful visual inspection by a knowledgeable arboriculturalist. Major defects are often apparent to a lay person. There are obvious benefits to looking at trees periodically so that significant hazards can be dealt with in a timely way.

The Occupiers' Liability Acts (1957 and 1984) and Liability Act (Scotland) 1960 require that premises do not endanger people. Trees are part of a property or premises and they must be properly considered so that people are not put at unacceptable risk. This applies equally to public areas and local authorities. This 'duty of care' increases as the size of the landholding increases. Thus, a cottage owner, while still responsible for the trees growing on the land, is not expected to have the same degree of competence as an urban authority or a leisure park, for instance.

Any urban tree manager is likely to be faced with the following questions when seeking to plan a monitoring and inspection programme:

- What is the objective of each tree inspection?
- Which trees should be inspected?
- Who should inspect trees?
- How should trees be inspected?
- How often should trees be inspected?
- What kit is needed to inspect trees?

Tree inspections do not take the place of tree management. The duty of care has not been met just because a tree survey or inspection has taken place. Once an inspection has been done, and the information is safely tucked up in the filing cabinet, it is important that the work identified is carried out. Failure at this point will result in loss of value of the tree resource, missed opportunities to act in a timely way to improve situations and a rising chance of an accident leading to damage or injury. If a record of an unsafe tree is held in an office and not acted on, the responsible manager is in a risky position!

WHAT IS THE DIFFERENCE BETWEEN A SURVEY AND AN INSPECTION?

- A survey is a general, overall assessment of a group of things, giving basic information about them. It is defined in the dictionary as 'a general view, casting one's eyes or mind over something'. An inspection is a more focused activity, taking one item of a group at a time and examining it.
- In arboriculture these terms often get jumbled up; surveys may involve detailed data collection of individual items and inspections can be cursory assessments of tree populations. So, let's not get hung up about the terms too much. Make sure the reasons for carrying out any survey or inspection are clear and tailor the process of collecting that information to those objectives.

You can tell a lot about a tree's condition by observing it as you approach.

WHAT IS THE OBJECTIVE OF EACH TREE INSPECTION?

Before any tree inspection or survey is carried out, inspectors and managers need to know what information to collect and why. Collecting information takes time and costs money. Local authorities may need particular information about trees to include in an environmental programme, detailing the contribution of trees to urban living, for instance, or the amount of carbon they have fixed, or are expected to fix, in their woody structure. Alternatively, a landowner may require a basic assessment of what work is needed over the next twelve months to limit risks to employees, guests and neighbours.

WHICH TREES SHOULD BE INSPECTED?

In the case of city-wide inventories there may be clear guidance provided in the objectives of the tree managers. Where safety is the primary driving force, the risk of harm from a tree becomes a deciding factor and the rest of this section concentrates on this approach to tree surveys.

Risk of harm from a tree is dependent upon three factors: the size of the weakest part of a tree, or the most likely part to collapse; the likelihood that it will fail; and the frequency of use of the area around it. That is, how likely is it that the failure of the tree will lead to damage or injury? Trees growing in the middle of a wood, for example, with no buildings or paths nearby and no easy access for people and vehicles, are very unlikely to hurt anybody, so do not warrant spending money on to inspect. Some trees on a site will therefore receive more attention than others and some may not need to be inspected at all.

The Health and Safety Executive has identified the minimum tree survey that complies with the safety guidance in the *Management of Health and Safety at Work Regulations 1999* and *Approved Code of Practice* (guidance is also contained in HSG 65 *Successful Health and Safety Management* and INDG 163 *Five Steps to Risk Assessment*). Following this guidance, summarized below, is likely to lead to compliance with legal duties of care, but as there are many other reasons for carrying out tree inspections it represents only the core of what is involved in a tree inspection.

An overall assessment of trees should be made to place them in one of two zones – where there is frequent access to trees and where there is no frequent access to trees. Trees in the frequently visited zone should be given a quick visual check for obvious signs of instability, carried out by a person with a working knowledge of trees and their defects, but who need not be an arboricultural specialist. Brief records should be kept of when a zone or any individual tree has been checked or inspected, including details of defects found and action taken. Where problems are identified that cannot be easily remedied, there should be a system for calling in specialist assistance.

The recording system should also enable inspectors or others to report damage to trees, such as vehicle collisions, and prompt checks following potentially damaging activities such as work by the utilities in the vicinity of trees, or severe gales. Where trees have serious structural faults but are to be retained, a specific assessment for each tree and specific management measures are likely to be appropriate. Any arboricultural work required should be carried out by a competent arboriculturalist.

Inspection of individual trees is only expected to be necessary where a tree is in, or adjacent to, an area of high public use, or has structural faults that are likely to make it unstable but a decision has been made to retain the tree. It is important that monitoring arrangements are carried out.

WHO SHOULD INSPECT TREES?

Any tree inspector must be fully qualified to meet the standards of the inspection. So, for example, an inspector entrusted with checking the effectiveness of tree stakes and ties of a young tree needs to know far less to do the job well than an inspector asked to determine the risk to passers-by of a large tree at a busy road junction. However, even an untrained tree owner is able to look at a tree with some degree of understanding and it is useful for non-specialist surveys to be done as an initial assessment, for instance after floods or storms, and to recognize where there is an obvious problem that needs further investigation by a more experienced inspector.

The Arboricultural Association and Lantra run two levels of courses for tree inspectors that recognize this range of skills. The first level of training is for inspectors from non-arboricultural backgrounds who need to be able to recognize major problems to trees from ground level and be able to report them to a line manager. The second-level Arboricultural Association and Lantra course is for competent arboriculturalists to enable them to identify defects and specify remedial works.

HOW SHOULD TREES BE INSPECTED?

Clearly, any tree inspection should be made for specific reasons that will determine what information is collected and what actions are appropriate. Safety will always be a major factor in tree inspections in urban areas because of

Decaying branches overhanging the highway should be removed to safeguard the public.

the potential of harm from trees that fall over or break up. So how do you carry out a tree safety inspection?

Landowners with no specialist knowledge of trees still have a duty of care not to put people and property at unacceptable risk, meaning that periodic assessments of trees are necessary. In the case of a householder with a couple of trees in the back garden or, for an estate owner with trees covering a whole area, it is likely to lead to someone standing at the base of each tree and scratching their head, wondering whether it is dangerous or not.

Step One: Zoning

Zoning is a primary consideration, in that trees growing where there are no people or structures, and no likelihood of damage, will not need inspection, whereas boundary trees, those near roads and buildings, or a footpath within or beside a site will definitely need checking.

A first step in identifying trees for inspection is to divide the area into zones of varying potential harm on the basis of frequency of use. Low harm potential is likely where areas are not used much. A high harm potential would be where frequency of use is significant. A middle, moderate zone may also be appropriate. Once a site or area is zoned, decisions can be made about how frequently inspections are needed and what level of competency is needed by the tree inspectors.

Step Two: Owner Inspection

Landowners must have some idea of what to look for when checking their trees. If damage is done and an insurance claim made, or a court case results, it is important that a landowner can show clearly that the duty of care has been met, that is, by acting responsibly, with common sense and a modicum of knowledge.

That knowledge extends to being able to identify obvious signs and symptoms of danger in trees. These include spotting large dead or hanging branches, cavities and decay on stems or large branches and instability at the base or in the root zone. Tree owners are also expected to be able to notice when something changes, for example the angle of the trunk, or the colour or density of the foliage. The action taken, following identification of these defects, is either to get a contractor to remove the hazard, or to get a consultant to investigate the matter further.

Step Three: Inspection by Knowledgeable Persons

For inspectors with some arboricultural knowledge, considerations of safety involve looking at a tree to identify the risks of it, or parts of it, falling over. Any safety inspection attempt to identify weak parts of the tree considers the likelihood of failure and to this adds the chances of damage or injury in the event of failure.

The principle of Visual Tree Assessment introduced by Mattheck and Breloer is now well established in arboriculture and Mattheck provides more details about the interpretation of the structure of trees in the *Updated Field Guide for Visual Tree Assessment*. A structured approach to tree inspection then would begin by observing the tree even while approaching it. Consider the surroundings. What is the ground-cover vegetation and what can you tell about the ground conditions? What does the tree look like from a distance? Does it look natural and healthy? When you know what a typical, healthy tree looks like, viewing a tree from distance can be helpful in picking up a general impression of whether it is thriving or not.

From a distance you are looking at the angle of the trunk, if it is visible, the colour and density of the crown and the presence of dead branches jutting out beyond living foliage. As you get closer to the tree, it will be possible to see the trunk more clearly and to gauge whether it is at a natural angle and also the main, scaffold branches will be seen more clearly. The ground around the base of the tree is always worth looking at as you approach the trunk. Are there signs of recent upheavals or excavations? Does it look like the roots can spread out unhindered, or is there so much concrete present that root growth must be restricted?

Once at the tree, information can be collected about roots, the trunk and the branches. Take time to look at each element of possible damage separately, for example: signs and symptoms of disease, pests or non-living agents such as wind or traffic; unusual growth patterns such as swellings and concave areas and sharp angles; extraneous materials such as timber nailed onto branches, washing lines wrapped around the trunk or railings being engulfed by the base. Then consider the leaves. These should be of typical size, colour and shape (so again, this means you need to spend time learning what the healthy trees look like). Look at any shoots that are within reach. What is the length of this year's shoot? How does it compare with shoots from previous years, further back along the branch?

All these observations should be recorded on a prepared, standard form, based on the objectives of the inspection. This is useful as an aid to step through all the major information that's needed, preventing important details being forgotten. With the tree information it is also important to record the date and weather, the location of the trees and the client's name.

As you look at the tree and record the information you may form an idea of what is happening to the tree. But this is the next step. Do not worry too much initially about understanding the signs and symptoms, just observe and record. When you have written down all that you can, stop and think. Why does the tree look like this? Do your observations suggest a problem and a reason for it?

If a survey is being carried out for reasons other than health and safety, such as an inventory of the tree resource in an area, it is still important to consider the health of the tree and take safety into account, but it may also be necessary to consider other factors, such as

Broken and hanging branches are a safety issue if people or property may be underneath.

the size of the tree, how much shade the tree casts, whether it obstructs a view or any other parameters specific to the survey objectives. If the age of the tree is needed, perhaps in relation to the age of an area tree preservation order or to help with comparing it to the age of nearby structures, the diameter at 1.5m (5ft) above ground level should be measured.

An important part of a tree safety survey is to consider the frequency of later inspections. Any assessment of tree safety is limited to the date on which it is made. Trees change over time and no one knows when the next big wind, flood, pest or vandalism attack is coming. This means that any prediction about what may happen to the tree becomes less reliable over time, therefore someone will need to come back and check at some point. So, recommend a timescale for this revisit.

WHEN INSPECTING TREES TO DIAGNOSE HEALTH PROBLEMS THE FOLLOWING MNEMONIC CAN BE USEFUL IN DIRECTING ATTENTION TO ALL THE NECESSARY AREAS TO INVESTIGATE (FROM STROUTS AND WINTER, 1994)

D distribution of damage

I identity of tree

S site type, conditions and changes

E environmental conditions and other changes (weather, chemicals etc)

A age of tree now and/or at planting

S symptoms

E extraneous matter (eg insects, fungi, exudates, odours)

D dead bark, discoloured wood – check by cutting

T time of onset of damage and its progress with time

R root condition

E evidence – review before making a diagnosis

E extra evidence – seek, where possible, in order to confirm diagnosis

If the trees are difficult to get at, this is important information to include on the record. You cannot state that a tree appears stable if you have not been able to assess the ground conditions at its base. It may be sufficient to record what you see from a distance, but some inspection objectives will require the ground around a tree to be cleared in order to get the job done. Recommendations will often involve return inspections by someone with more knowledge or more specialized equipment.

Step Four: Specialist Inspection

Once trees have been looked at, either by the landowner or their knowledgeable agent, there may well be some issues that are ambiguous or of concern. At this point, an arboricultural specialist is needed. The greater insight they bring may resolve the matter and suggest appropriate action, or specialist equipment, such as the Picus Sonic Tomography, IML Resistrograph, TreeRadar or thermal imagery, may be needed to provide more information on which to base a judgement.

Step Five: Keeping Records

Keep written records of inspections, assessments and actions. Include a plan of the trees inspected and take photos. This provides clear, consultable information, which builds up over time into a valuable resource for tree management.

Subsidence is a Special Case

Tree inspections for the purpose of assessing factors involved in subsidence damage are a special case. Here, the condition and age of the tree are compared to the age of the damaged building, plus other factors are considered, such as the type of subsoil, presence of other vegetation, tree roots found close to the area of damage and the pattern and character of that damage. Only some of these factors are tree-related, so the interpretation of the survey information is a team effort between an arboriculturalist and engineer and possibly other specialists.

HOW OFTEN SHOULD TREES BE INSPECTED?

The current guidance for highways authorities is that all highways trees should be inspected at least once every five years (DoT, 2005). This is the clearest specification for inspection frequency and it obviously relates to trees along the edges of roads that have the potential to do significant harm if they collapse. There is little other guidance to use.

Large landowners or local authorities will need to balance the frequency of tree inspections with limited budgets and the resulting inspection regime may reflect their financial constraints more than a biological optimum. But although lack of money can be argued as a mitigating factor, if things go wrong and a landowner ends up in court, it cannot take away their duty of care and the responsibility to manage trees and prevent them being an unacceptable risk to others.

Given that trees are dynamic, changing structures subject to variable environmental conditions, it is clear that a lot can happen to trees over a five-year period. In fact, it is difficult to anticipate how a tree will be affected by climate and other planned or unplanned changes to its surroundings over even a twelve-month period. So there seem to be two questions to ask: what is a realistic return period for inspections given the financial resources of the landowner? And what return period is likely to be effective in picking up problems in tree health and safety issues before an accident happens?

For small-sized properties, it would seem reasonable for tree owners to make a quick, visual check on an annual basis, then to call in someone with specialist knowledge if anything of concern is noticed. It is also prudent for tree owners to check their trees after extreme climatic events, such as storms or floods. These are recognized situations that could affect the stability and safety of trees.

As the size of landholding increases, an annual survey becomes more difficult to programme, therefore a judgement will need to be made about what is reasonable and practicable and what is required to comply with the duty of care. On large landholdings, a more knowledgeable tree inspector will be required, whose expertise can help to guide appropriate inspection frequencies.

Once an inspection regime is in place it should be fairly straightforward to keep it working smoothly, with all trees near to people and property being surveyed on a regular basis and all work identified being carried out in a reasonable timescale. But what happens if no tree inspections have ever been done? Maybe the landowner has changed, or the liability for trees has only recently been recognized. In this case, it is important for the landowner to obtain an understanding of the condition of the tree population as quickly as possible, so that a prudent decision can be made about what works are necessary and what money needs to be spent. Just setting up a five-year programme could leave dangerous trees hanging over a road or building for an unacceptable length of time.

In this situation, the best approach is to establish the areas or trees of highest risk. This is a refinement of the zoning stage of a tree inspection programme. The first priority must be to find those trees of highest risk and deal with them. So a team will need to be given the job of checking the whole area in as short a timescale as is practicable. A regular programme of tree inspections should also be started so that, over time, the whole area can be treated equally and effectively.

WHAT KIT IS NEEDED TO INSPECT TREES?

The first requirements are an enquiring mind and some knowledge of tree biology. Equipment cannot take the place of an inquisitive, knowledgeable inspector.

At its most basic, all a tree inspector needs is a pair of strong boots and a pencil and paper.

A tight, compression fork and included bark have led to this willow trunk splitting.

But there are some helpful things that can be added to this minimalist list:

- diameter tape – this is a primary parameter for determining tree age and characterizing the size of a tree; tree diameter is also useful when checking back with previous inspections to assess growth rate
- clinometer – for measuring the height of trees
- binoculars – to help to focus on potential weak points on the tree that are too far away to see clearly
- knife – very useful for poking around in small holes and bark
- rubber hammer – for tapping tree trunks; hollow areas can be identified by sound
- long stick or metal rod – useful for poking into hollows and cavities to determine depth
- spade or trowel – for digging around the base of a tree when assessing roots
- secateurs and loppers – often epicormic growths need to be removed to allow inspection, or adjacent plants need to be pruned back to get at the tree
- if samples of tree or soil are to be taken, a soil auger will be useful and polythene bags and possibly plastic containers will be needed; don't forget to take labels for the bags.

Safety of the inspector is important and personal protective equipment (PPE) will be needed along the following lines:

- strong shoes or boots, preferably steel toe-capped
- warm, rainproof coat; this is Britain!
- high-visibility jacket – most surveys involve working close to roads at some point
- mobile phone – to keep in contact with the office for operational purposes, but also for personal safety; someone in the office needs to be aware of you being on site, possibly alone, and accidents or unexpected events do happen
- on development sites, or where there is a recognized risk of falling objects, a hard hat is appropriate
- a first-aid kit is an essential part of an inspector's equipment.

Specific surveys, such as for irritant insects or where poisonous materials are expected, may require further protection such as gloves and goggles. In urban areas the risk of 'sharps', discarded needles, broken glass or similar items means that at all times a tree inspector should consider personal safety. Gloves may be needed; it is always wise not to shove an unprotected hand into an unexamined cavity. It is easy to forget such details, so make it part of your usual procedures on a site. PPE requirements can vary over time, on different inspections and in response to risk assessments, so this aspect should be regularly reviewed.

If an inspection requires climbing a ladder or climbing the tree, it is advisable that this should only be done by qualified persons and be done accompanied by a qualified colleague. My personal procedure is to take a tree surgeon with me when a climbing inspection is needed. The tree surgeon can put the rope in the tree safely and quickly, which makes my job easier. He or she can then monitor my progress within the crown and rescue me if I need it!

Obviously, with all these matters it is important only to use equipment for which you are trained and safety procedures should always be followed to avoid problems when untoward events occur. Public liability insurance is needed for all survey teams or inspectors, plus professional indemnity insurance will be needed if reports are to be issued to clients.

Specialist Equipment

Where a visual tree inspection is not able to recommend specific action due to lack of information or hidden problems are suspected, specialist equipment may be required to investigate the situation further.

An increment core can be used to look at the condition of the wood and also many trees have annual rings that can be used to find out their age.

Small bore drills, such as the IML Resistograph, can be used to test the strength of internal wood. The drill is bored into the wood and a stylus records the resistance to the drill on wax paper. Regions of sound wood show high resistance to the forward motion of the drill bit, but decayed areas show reduced resistance and the trace reacts accordingly.

Ultrasound systems, such as the Picus Tomograph, can be used to map internal decay or changes to the wood properties over a transverse section of a trunk, branch or root. Electrical impedance tomograms are an additional way of gaining information about the condition of wood. Ultrasound tomograms measure the time taken for sound waves to pass from a transmitting sensor to receiving sensors placed around the circumference of a stem. Electrical impedance tomograms show changes in wood characteristics because decay increases conductivity.

Ground penetrating radar can be used to map the position and depth of roots.

Resistograph microdrill.

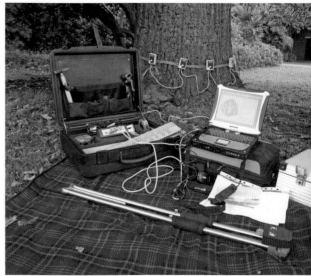

Picus tomography equipment.

Climbing inspections may be needed to inspect defects in the crown.

TreeRadar equipment.

Heat-sensing photography is being developed that can show areas of decay within trunks.

THE LONE INSPECTOR

Working alone may have some advantages, but it also has many limitations and requires careful thought so that the inspector is fully equipped to deal with whatever the job may throw at him or her. There is no actual legal problem regarding lone working, but a risk assessment should be made and the working method adjusted in the light of its findings. Lone working is not recommended when working near water, along railway lines, where there is a risk of violence, or if trees need to be climbed or inspected from a ladder.

Even if working alone, a worker should be part of a wider team and the team members need to stay in contact with each other. Information about the progress of the inspection should be reported at agreed times and the lone inspector should call in when knocking off so that the office knows all is well.

The Home Office provides information on personal safety on its website: www.homeoffice. gov.uk/documents/besafebesecure.pdf.

TREE VALUATION

Tree valuation is not a settled aspect of urban tree management. How to value trees in different circumstances and for different reasons has been a thorny issue for decades and further development and integration of the current options are needed in order for trees to be effectively factored into economic decisions in towns and cities. In today's world, if an asset or resource cannot be given a monetary value it is likely to be seen to have no value, or be impossible to compare with other assets that are easily valued in this way. Trees are difficult to value effectively, so easily lose out in any economic assessment. This ignores their real value and squanders the benefits they provide, or could provide.

The Royal Institution of Chartered Surveyors (RICS) has recently produced a guidance note for the valuation of amenity trees. This document highlights the difficulties in effective valuation; sets out the professional valuer's approach; and highlights the Helliwell, CTLA and CAVAT methods. The note focuses on tree values on individual sites and primarily in regard to land valuation.

Until recently, most assessment methods failed to include an element for social value. The CAVAT method attempts to include this factor and builds on the experience and resilience of the Helliwell and CTLA approaches. Also included below is the i-TREE approach to tree valuation from the USA and the TEMPO method of tree valuation in order to aid the decision regarding whether a tree preservation order is appropriate.

The Helliwell Method

This method was devised by independent consultant Rodney Helliwell and has been in use in the UK since 1967. The system aids urban tree planners and managers by considering the relative contribution of different trees and woods to the visual quality of the landscape. The value of a tree or woodland is calculated by multiplying a range of factors that are given numerical values. The final number is converted to a monetary value using an index-linked figure.

The method can be used to assess individual trees, groups or woodlands. The factors incorporated into the calculation for trees are:

- size of tree
- expected duration or life expectancy
- importance of the tree
- the presence of other trees in the vicinity
- the relation of the tree to its setting
- tree form.

The assessment for woodlands includes the following factors:

- size of woodland
- position in landscape
- viewing population
- presence of other trees and woods in the vicinity
- composition and structure of the woodland
- compatibility of the woodland with other land forms or features.

Some of the factors are fairly straightforward to apply, but others require considerable arboricultural knowledge.

Council of Tree and Landscape Appraisers (CTLA)

The CTLA method has developed from an original procedure from the USA. It values tree stock by using financial asset appraisal, estimating the potential market value by using methods that are based on a market approach, an income approach, or a cost approach.

Market approaches to tree valuation rely on the availability of sufficient information to compare the value of a property or an area with trees to an equivalent one without trees. An income approach uses market information about the value of the trees as income streams to a business, or as positive, measurable factors

that can be assessed in an economic appraisal.

However, the most common tree appraisal methods are examples of the cost approach: replacement cost method (RCM); trunk formula method (TFM); and cost of cure method (COC). These are all depreciated replacement cost approaches to value. They all recognize three depreciation factors: species; condition; and location. These factors modify an initial cost-estimate to characterize a particular subject tree. The RCM calculates how much it will cost to replace a lost tree. The TFM calculates a value per unit of trunk area, while the COC method works out the costs incurred in remedying a damaged tree. This approach to tree valuation is currently being modified to be more suited to UK conditions.

Capital Asset Value for Amenity Trees (CAVAT)

CAVAT is a system of valuation developed by Chris Neilan for use in the UK and adopted by the London Tree Officers Association (LTOA) for use on publicly owned or publicly important trees; either individuals or the whole urban forest stock of an authority. It calculates tree replacement cost per unit area and incorporates a social value element via the Community Tree Index. This index is derived from population density statistics, held by the Office for National Statistics, the principle being that tree values rise with the number of people that interact with them.

The baseline figure is translated into a monetary figure and the trunk diameter is then factored in to give a final value. The full method involves five steps and sets of key variables:

- Step 1: basic value (unit value factor × area of trunk at a height of 1.5m [5ft]).
- Step 2: Community Tree Index (CTI) value. This step factors in the population density of the neighbourhood, which is then discounted according to how accessible the tree is.
- Step 3: functional value. This is a reduction in

value in proportion to the health of the tree. It requires an assessment by an experienced arboriculturalist.

- Step 4: adjusted value. This is found by incorporating any relevant special amenity factors into the functional value.
- Step 5: full value. This is the last step and includes an assessment of the safe, useful life expectancy of the tree, reducing the tree value as the safe life expectancy shrinks from eighty years or more (100 per cent factor) to less than five years (10 per cent factor) and to zero if the tree cannot be retained safely.

The Quick Method enables the value of a public tree stock to be expressed as an index. The index would rise or fall as the quality and character of the tree stock changes over time and the tree manager uses the management information to decide on interventions and to track the change in value of the resource. The data required is kept to the minimum that allows this. Three factors only are used:

1. basic value; tree size is included by use of sixteen value bands that are updated regularly
2. the CTI value is needed
3. the tree's relative functional health and integrity are considered in 25 per cent gradations; this requires arboricultural expertise and training.

i-Tree

This is a software suite from the USDA Forest Service that provides analysis and benefits-assessment tools for urban forests. The i-Tree Tools are designed to help communities of all sizes to become involved with and improve urban forest management. The tools provide a method to quantify the structure of trees in a community and the environmental services that trees provide.

The i-Tree Tools were originally released in August 2006 and are under constant development. The tools rely on random sampling plots stratified by land use that give information about the tree resource and on local pollution and meteorological information. This data enables local, tangible ecosystem services to be quanitified and linked to urban forest management activities. The i-Tree software provides baseline data about individual trees or whole woodlands that can be used to demonstrate value and set priorities for more effective decision-making.

The i-Tree Tools are in the public domain and are freely accessible from http://www.itreetools. org. They are designed primarily for North American conditions, but work is currently being undertaken to make the tools more relevant to urban areas in the UK. Software programs are available that model the entire urban forest: i-Tree Eco; the local authority street tree population: i-Tree Streets; and there are also utility programs to help choose appropriate species for planting: i-Tree Species; and to assess storm damage: i-Tree Storm.

If this approach and the expertise developed in the USA can be harnessed for UK conditions, it could help enormously in quantifying our urban tree resource.

Tree Evaluation Method for Preservation Orders (TEMPO)

The TEMPO system was devised by Julian Forbes-Laird (then at CBA Trees) to provide a reliable means of assessing trees for tree preservation order (TPO) suitability. The method is divided into three parts: part one is an amenity assessment; part two an expediency assessment; and part three is a decision guide.

The amenity assessment considers four factors:

- tree condition
- remaining longevity
- relative public visibility
- other factors.

If the numerical score from part one reaches seven points or more, the part two assessment is undertaken. The expediency assessment awards points based on the level of identified threat to the tree. If the accumulated score for parts one and two reaches eleven to thirteen points, it is appropriate to consider protecting the tree. If the score exceeds thirteen a TPO is recommended.

REFERENCES

Barrell J., *The Emerging Duty of Care in England Relating to Trees: A Practitioner's Perspective*, paper submitted to RICS for publication (RICS, 2010).

CTLA, *Guide for Plant Appraisal*, 9th edn (ISA, 2009).

Cullen S., *Tree Appraisal: What Is the Trunk Formula Method?* (9th edn) (Tree Appraisal Theory and Practice seminar, Bournemouth, 2002, 2000).

DoT, *Well-Maintained Highways – Code of Practice for Highway Maintenance Management* (TSO, 2005).

Fay N., Dowson D., Helliwell R., *Tree Surveys: A Guide to Good Practice* (Arboricultural Association, 2005).

Forbes-Laird J., *Tree Evaluation Method for Preservation Orders* (FLAC, 2009).

HSE, *HSG 65 Successful Health and Safety Management* (HSE, 1997).

HSE, *Management of Health and Safety at Work Regulations 1999 Approved Code of Practice* (HMSO, 1997).

HSE, *A Short Guide to the Personal Protective Equipment at Work Regulations 1992*, INDG 174 (rev1) 08/05 (TSO, 2005).

HSE, INDG *163 Five Steps to Risk Assessment* (HSE, 2006).

HSE, *Management of the Risk from Falling Trees*, SIM/01/2007/05 (HSE, 2007).

HSE, *Tree Climbing Operations*, AFAG leaflet 401 (HSE, 2009).

Lonsdale D., *Principles of Tree Hazard Assessment and Management* (TSO, 2001).

LTOA, *A Risk Limitation Strategy for Tree Root Claims*, 3rd edn (LTOA, 2008).

Matheny N.P. and Clark J.R., *A Photographic Guide to the Evaluation of Hazard Trees in Urban Areas*, 2nd edn (ISA, 1994).

Mattheck C., *Updated Field Guide for Visual Tree Assessment* (Forschungszentrum Karlsruhe GmbH, 2007).

Mattheck C. and Breloer H., *The Body Language of Trees* (HMSO, 1994).

Neilan C., *Capital Asset Value for Amenity Trees (CAVAT)* (LTOA, 2008).

Sacre K., *i-Tree Abroad and Now at Home* (Essential Arb, 2010).

Schwarze F.W.M.R., *Diagnosis and Prognosis of the Development of Wood Decay in Urban Trees* (Enspec, 2008).

Smiley E.T., Fraedrich B.R., *A Guide to Managing Trees on College and Corporate Campuses* (Bartlett Tree Research Laboratories, undated).

Strouts R.G. and Winter T.G., *Diagnosis of Ill-Health in Trees* (HMSO, 1994).

chapter seven

Urban Tree Management in Public Places

Urban areas include many publicly owned sites containing structures and infrastructure where people are invited or expected to congregate. In the best of these areas trees grow to soften the hard effects of a built landscape. Large parts of urban areas are used for travel and access, for example roads, paths, driveways and streets. At the edge of these transport arteries are roundabouts, verges and banks. In these areas are road signs, lamp columns, underground services and, increasingly, closed-circuit television cameras.

Seen from almost any angle, urban areas are a tangle of concrete, glass and tarmac covering the land. Large tracts of countryside are now covered by the spreading of towns and cities and a transition can be seen at the edges where there are still fields, hedgerows and trees. New developments create stark, hard landscapes with little mature vegetation, but, over time, plants soften the edges and bring hints of nature into the man-made setting. However, by definition urbanization reduces the naturalness of an area. Town centres and industrial estates have often been intensively developed over decades, if not centuries, and the incorporation of trees, plants and water is now recognized as an essential element of good design that is required to create a healthy urban environment in which people can live and work. The management of trees in these intense

conditions requires a clear understanding of what trees need to survive and thrive, but also it has to be guided by the needs and wishes of residents and users.

THE PURPOSE AND VALUE OF TREES IN PUBLIC PLACES

The role and value of trees in urban settings are becoming more accepted and recognized as our urban areas expand or require renewal. Typical roles that are fulfilled by urban trees include screening, framing, softening and shelter. Trees are valued for their naturalness, historic or cultural connections and their physical presence and character. Trees provide valuable environmental services by absorbing and releasing water, absorbing pollution and by blocking wind or sun. Emotional and psychological benefits include reducing stress, improving mental health, encouraging recovery and enhancing well-being.

While individual trees provide benefits, these are magnified in larger areas, therefore parks and open spaces, already important in urban areas, are crucial in providing the services and functions that are needed. Changes in climate and the increasing understanding of urban conditions are emphasizing the need to nurture and develop trees within towns and cities. Trees are part of the urban life-support system and this role is likely to become more pronounced throughout the twenty-first century.

STREETS AND HIGHWAYS

Urban roads come in many shapes and sizes, from a residential cul-de-sac to a motorway. Clearly, the opportunities for establishing thriving trees vary in these differing situations. At one end of the spectrum a narrow residential road may have no trees along it and offer no realistic possibility of planting any, whereas a wide soil bank at the side of a motorway gives more opportunities for planting and will be naturally colonized by trees anyway unless management involves regular mowing.

Roads and Streets With Trees

Many urban roads have trees growing near their edges. Appropriate management depends on the species and the space around them, both above and below ground. The above-ground management is the easy part, because it is likely only to involve organization and implementation by one department of a local authority. Large-growing trees, such as limes, planes or maples, may well need to be contained within a limited volume of space to avoid damage and inconvenience at an unacceptable level to residents and adjacent features. Small-sized trees, such as crab apples, cherries and rowans, are likely to be less troublesome to neighbours. Even when mature, such trees are unlikely to be much taller than 10m (33ft).

The above-ground space available for a tree to grow is dependent upon how close it is to the road and to adjacent buildings. The Highways Act 1980, section 154, gives each highways authority the duty to keep the highway clear, but it does not stipulate a precise height requirement. The minimum clearance above a road is, therefore, interpreted generally to be around 5.5m (18ft) (this varies slightly between highways authorities, but is guided by the minimum maintenance clearance above free-standing temporary structures stipulated in TD27/05 as 5.41m [The Highways Agency, 2007]), above a footpath it is around 2.3m (7.5ft). This guidance in the *Design Manual for Roads and Bridges* (*DMRB*) uses a maximum eye height of 2.2m (7.2ft) for cyclists and 2.7m (9ft) for horse riders, so height clearances of cycle paths and bridleways need to be greater than

Street trees, like these in Brighton, provide many benefits to residents, visitors and shoppers.

Street trees filling the space above the road.

for pedestrian-only footpaths (*The Geometric Design of Pedestrian, Cycle and Equestrian Routes*, TA90/05). These clearances are needed to avoid damage to passing vehicles and to enable pedestrians to walk unhindered down a road, even in the rain when using an umbrella. The clearances extend by around 300mm (12in) to the sides as well as directly above a road or path in order to avoid sideways encroachment. This is a crucial concept to bear in mind on roads with significant cambers, where the road surface curves down from the centre to the gutters at the pavement edge. This camber can tilt high-sided vehicles towards the edge of the road and therefore toward trees. A 300mm (12in) clearance from the edge of the road does not make much difference if a bus, of around 4.5m (15ft) tall, leans over. This equates to a lean of only just over 3 degrees. This camber should be kept in mind when planning new tree planting to avoid such problems in future.

But there are thousands of existing street trees that are being bashed, in danger of being bashed, or being savagely pruned to

Buses and trees do not mix. Either keep the bus away or prune back the tree.

avoid being bashed. Many bus companies use a modified, open-topped double-decker bus to prune overhanging branches on bus routes and they may identify large branches or tree stems that are just too much of a risk to buses and lorries. These trees will inevitably be pruned to meet the imperative of keeping the highway safe. But there are many, less extreme situations where local knowledge and good communication can help to reach a compromise that safeguards the tree features and alerts the bus drivers to the practicalities of their routes. To help achieve this, branches or stems can be painted or marked, kerbs can be built out to keep vehicles at a greater distance, or the road surface, and possibly bus stops, can be built up to reduce surface camber. All of these measures require a positive, problem-solving approach between the tree managers and the highways authority managers.

Opportunities for Growing Trees on Roundabouts and Approach Roads

Highways engineers can be reluctant to accept tree planting on or near to junctions, but the guidance in the *DMRB* does not exclude trees in all instances; in fact, it encourages tree planting for screening in some situations. Along

ROUNDABOUT DIAMETER AND VISIBILITY DISTANCES	
Roundabout diameter (m)	Visibility distance (m)
<40	Whole junction
40–60	40
60–100	50
>100	70

with the central island of roundabouts the approaches, which may include splitter islands between carriageways, should be considered.

The principle used by engineers to determine the safety of roundabouts and junctions is based on visibility. Along roads and away from junctions, the principle is the desirable minimum stopping sight distance (SSD). At roundabouts, the principle dictating where not to plant is the visibility distance, which is related to the diameter of the roundabout junction.

Central islands less than 10m (33ft) in diameter are unlikely to be suitable for any tree planting (section 8.41 of TD16/07 [The Highways Agency, 2007]). The volume of soil is likely to be limiting for the tree and maintenance could be difficult,

A roundabout of 58m (190ft) diameter with 40m (130ft) sight lines added. 1. Sight line 40m in front of the give way line, nearside carriageway. 2. Sight line 40m to right of the give way line, outside carriageway. 3. Sight line 40m to right from outside carriageway, 15m back from give way sign.

with no room for parking. This would mean that any maintenance team would need to run across the road with irrigation equipment, pruning tools and other materials.

On larger roundabouts, there are opportunities for planting to increase the visibility of the junction, or to blend in with the surrounding landscape. Screening of traffic on the opposite side of the roundabout can also be a positive factor. Planting provides a positive background to chevrons that may be painted onto the sides of the central island.

But in all cases, the outer 2m (6.5ft) of the roundabout will need to be kept clear of all tall vegetation in order to maintain good visibility for motorists. Visibility splays around and across the central island and splitter islands are taken from the give-way line or a point 15m (49ft) back, to represent the approach to the junction. Looking to the right, on approaching the junction and at the give-way line, the motorist in the outside lane must be able to see the carriageway for a distance relative to the diameter of the junction (see the table). Looking forward, the motorist at the give-way line of the nearside lane must be able to see the carriageway ahead to a distance indicated by the diameter of the junction.

The visibility splay limits the area of the central island or splitter islands available for planting, but it does show that there are significant opportunities for planting on larger junctions where the presence of trees can be presented in a positive light. Clearly, where islands are wider, there is more scope for taller trees and denser planting of shrubs and trees. But where roundabouts on high-speed approaches are not protected by high edges or other strong barriers trees are discouraged.

On approaches to a roundabout, excessive visibility is as unhelpful as very limited visibility, leading to high entry speeds at junctions. On dual carriageway approaches with speed limits above 40mph, traffic can be slowed by limiting visibility to the right until 15m (49ft) from the give-way line using tree screening, which should be at least 2m (6.5ft) tall.

Other practical issues to face when considering planting on a central island include knowing what underground services are present. These can severely restrict the available soil for root growth and require access for future maintenance. Soil volume and quality is a major consideration and unless an appropriate volume of good soil can be provided any tree planting will be a speculative activity. Future maintenance of the trees is an issue and access for the maintenance team should be provided. So, provide enough soil for the number of trees and other vegetation, plant them where the highways engineers will not view them as a major obstruction and include safe access for maintenance. That way there is every chance the planting scheme will be a success.

Away from roundabouts the principle of SSD informs decisions taken by highways engineers about safe visibility splays around bends or corners and other features of the road. So where traffic is travelling faster, vehicles take longer to respond to hazards and so need greater visibility in order to take appropriate action in good time.

Any tree planting on bends therefore needs to be kept back from road edges. For existing trees there may be some pressure for their removal if they are deemed to be obstructing the safety of the highway. However, there is some flexibility within the *DMRB* where environmental features are seen to be of value.

HIGHWAYS DESIGN SPEEDS

Design speed	kph	120	100	85	70	60	50
	mph	75	62	53	43	37	37
Stopping sight distance							
Desirable minimum	m	295	215	160	120	90	70
	ft	968	705	525	394	295	230
One step below desirable minimum	m	215	160	120	90	70	50
	ft	705	525	394	295	230	164

The SSD distances can be relaxed slightly, therefore the engineer or designer should be encouraged to consider the costs and benefits of trees as part of his or her assessment. Any assessment of this kind should be recorded so that there is a clear reference document to explain decisions. Such relaxations to the desirable minimum distances are not permitted on the approaches to junctions.

Banks, Verges and Woodland Edges

The sides of roads, including banks, slopes and cuttings, are often good locations for tree planting. In most places trees grow on this land regardless of planting schemes – it is a reassuring sign of nature at work. But nature does not comply with road safety regulations and as vegetation grows it can impinge on the highway in a number of ways.

The requirement for highways to be maintained clear of obstruction has already been discussed, but tree growth can affect road users by obscuring signs and signals and by obstructions from maintenance vehicles. Roadside maintenance should be planned to be periodic and have clear parameters to cut back vegetation to allow safe use of the highway until the next programmed pruning operation. So, for example, if a section of road is to be pruned back from the road, the amount of pruning must be sufficient that regrowth does not become a significant problem before the next visit. If visits are programmed at three-yearly intervals, it makes sense to cut branches and young stems back to give, say, a four-year clearance from sight lines or the road edge.

In these operations the correct equipment for cutting should be used. It is not unusual to see roadside trees and shrubs flailed back into large-dimension branches and so left with tattered bark and major wounds that can severely weaken them. Tractor-mounted side arm flails should be used only to prune back vegetation that is no older than two years. Branches older than two years should be pruned

using reciprocating blade-cutting machinery. Flails are therefore best used for hedges along roadsides that need annual trimming, or for interim operations to keep vegetation in check before it can be properly dealt with.

Linear Tree Belts

Linear tree belts are found where trees grow at the side of roads, or they may be planted along narrow strips, for instance along a central reservation between carriageways. Any maintenance to such features is likely to involve careful traffic management. The space and ultimate size of trees should be considered at an early stage in the management process. Large trees that will extend over the road and beyond the space available will be a continual and increasing drain on maintenance resources. Planting that is too dense will quickly lead to the need to reduce the numbers of trees, with problems arising if this thinning is delayed. Management should aim to match the space available to the number and size of the trees in the tree group.

Avenues

In cities, avenues tend to emphasize, soften or contrast with adjacent buildings. They can mark an important route into or out of town, may be used to add grandeur to a town centre or to provide some link between nearby parks. They may even be sited carefully to screen or minimize a nearby, unattractive feature. This sounds like standard management issues for individual trees and groups in towns. However, in some respects the management of avenues is different.

Avenue Management

Trees come and trees go, but if an avenue is seen to be an important feature it will need some careful thought to ensure there is a clear understanding of its value and the challenges it faces. There are three steps to effective avenue management:

A veteran avenue of beech trees, Kingston Lacy, Dorset.

- understanding the value of the avenue and then considering the reasons for management. Why is it there and what role does it or could it perform?
- finding out the condition of the resource. That is, what is there and where is it? There is, clearly, some overlap with these two objectives.
- formulating an effective plan to deliver the benefits that the avenue provides or is intended to provide. This includes considering options for moving from one generation of trees to the next. How will the avenue be preserved over long timescales?

There are times when a long-term view is not possible, due to lack of expertise, finance or political will. But, especially with the management of avenues, it is vital that a long-term view is taken and the fate of the avenue not left to short-term inexperience or caprice. Short-term management may have to grapple with what to do next, which may range across the whole spectrum from 'fell the lot' to 'do nothing'. There are likely to be times when either the 'fell' or 'do nothing' option is appropriate, but the safer short-term response is to do nothing, as it is less certain to destroy the resource.

Managing Neglected Avenues

Formal avenues have usually been established as single rows of trees along one or both sides of a road, or they may be double lines or treble lines as in a quincunx pattern. The more intricate patterns take up more land and are likely to have been used in Britain for avenues connected to a country house. They are thus

A SIMPLE QUINCUNX AVENUE PLANTING PATTERN (Xs indicate planting positions)
X   X   X   X   X   X
X   X   X   X   X   X
X   X   X   X   X   X
Road or drive
X   X   X   X   X   X
X   X   X   X   X   X
X   X   X   X   X   X

more likely to be encountered in an urban situation where a city has engulfed a country estate.

Avenues rely on uniformity and unity for their effect, so removing and planting trees can seriously alter their impact. Some alternatives to consider include:

- felling every second or third tree to maintain some uniformity
- retaining every tree until it is moribund or unsafe, thereby extending the life of individual trees, but reducing the unity of the whole
- removing groups of trees to provide space for new planting, which changes the avenue structure by breaking up the regular spacing of the original pattern
- retaining the whole avenue and planting additional lines of trees outside the original ones, which will give a different effect from the original avenue and requires lots of space
- felling every tree in the avenue once the overall effect has been diminished by disease, variability or tree removals, treating the whole feature as a single element and replanting to begin the process of avenue creation again.

Informal avenues are less easily spoiled by *ad hoc* removal and replacement planting because of the lower level of uniformity and unity. Single lines of trees may be managed separately and only considered as part of the larger whole if pruning is intended to formalize the shape of the tree crowns, or if there is a threat to the whole feature (which could be an aggressive pest or a road-widening scheme). The value of the avenue may still be very high and retention still a desirable objective.

An informal avenue, especially if it includes a range of species of tree, will lead to a greater wildlife diversity than a single-species, single-aged avenue. Rural avenues, that have been left some distance from surrounding features by changes to roads, buildings, society or agricultural practices, are likely to develop an increasing value as wildlife habitats over time. This can make management more complicated as the biodiversity objectives may place stringent controls on what is allowed to happen to the feature.

Planting New Avenues

New avenues should be planted with replanting and replacement in mind. The crossover between generations is so vital when managing avenues that it must be considered at the start of the endeavour, or it will remain an uncertainty, undermining future efforts at management.

The space available for the new avenue must be carefully considered. Below ground the rooting space must be of sufficient volume and quality to allow the trees to grow to the stature that enables them to provide the benefits for which they are being planted. Lack of space below ground means that the trees will either die or stay small; either way, the objective of the avenue will not be realized. There are products and materials that make this a realistic expectation now, whereas a couple of decades ago the options were far fewer. Excavation of a whole strip of footpath beside a road may be worthwhile to provide a continuous trench that can be backfilled with good-quality topsoil for the benefit of all the trees planted into it.

The space above ground is equally crucial. What size are the trees planned to reach? How wide and high will the crowns grow? How will the trees be kept to an acceptable size? All of these issues need to be considered at the start, so that the right avenue, consisting of the right species, planted in the right way, in the right places, will result.

Scrub

Land that is disturbed and then left alone will often become scrubland, and begin to be occupied by short-lived, fast-growing tree and shrub species such as birch, ash and hawthorn trees and shrubs such as bramble, gorse and bracken. The trees that establish themselves early on after a disturbance such as fire or development are called pioneer species. These trees are usually shade-intolerant and short-lived and, because their seedlings cannot survive the shade the parent trees cast, they are eventually replaced by species whose seedlings are more shade-tolerant.

Areas that can be left to develop naturally as scrub can become valuable wildlife habitats. They are also vulnerable to fire, which can be used as a tool for management; controlled burning can be used to renew the scrub species and to remove older, woodier vegetation that is a higher fire risk. Obviously, fire-prone species like gorse have to be kept well back from highways.

Amenity Tree Planting

Suburban and urban areas often include soft landscaping with opportunities for tree planting. In this way, redeveloped areas are softened and the tree population increased. But these areas suffer from problems such as vandalism, neglect, litter accumulation and road salt. Also, being close to roads, they must still be managed to avoid motorist visibility problems.

Personal security is also a growing problem in the perception of urban dwellers. Areas of dense shrubs or thickets of trees are seen as inherently dangerous places. Managing this issue requires a lot of sensitivity and diplomacy to find a management regime that balances the security issues with the other benefits provided by trees. A further constraint on space for trees in towns is the increasing use of closed circuit television (CCTV) set up by concerned authorities and businesses. It is estimated that there are 1.5 million CCTV cameras in the UK (Boddy, 2011). Each camera demands clear lines of sight that can lead to drastic tree pruning. The best way to minimise tree pruning or removal is to include trees at an early stage of planning for any CCTV project and to be aware of possible projects when planning new tree planting, positioning young trees away from known or likely sight lines. The CCTV code of practice states that cameras positioned in winter should take into account growth of spring and summer foliage (section 6, CCTV Code of Practice, 2008). So, CCTV managers have no excuse for badly positioned cameras and valuable trees should be retained and the cameras relocated where there are conflicts.

Establishing New Trees along Urban Streets

The *Well-Maintained Highways* code of practice (Roads Liaison Group, 2005) recognizes the contribution of highways trees to urban areas and it advocates a close cooperation between highway engineers and arboriculturalists, landscape architects and urban designers, so that existing and new trees can reflect the history, architecture and tradition of places. It recognizes that trees are not appropriate in every situation, but it does reinforce the need for highways engineers to see trees as more than obstacles.

When planting trees in urban streets, the whole life of the tree and its consequent maintenance needs should be considered. What is the point of planting a tree beneath an overhead power cable or in the way of an important CCTV camera? It is a wasted investment from the start. Areas of high vandalism are a similar challenge, but any urban tree planter needs a

persistent streak, so, if a balanced assessment of the situation suggests that tree planting is a good idea, the planter should be prepared to replant and replant again to achieve the planned objective. Vandalism can be periodic and it may be short-lived if the vandal group moves on after a few years.

Some strategies to overcome vandalism include using smaller trees or larger trees, using stronger trunk guards and resiting the trees to avoid vandalism hotspots. Smaller trees growing together in groups may not need any other protection. They avoid trouble by not being regarded as worthy of vandalizing, but they do need open space to be planted in this way. Larger trees are more difficult to

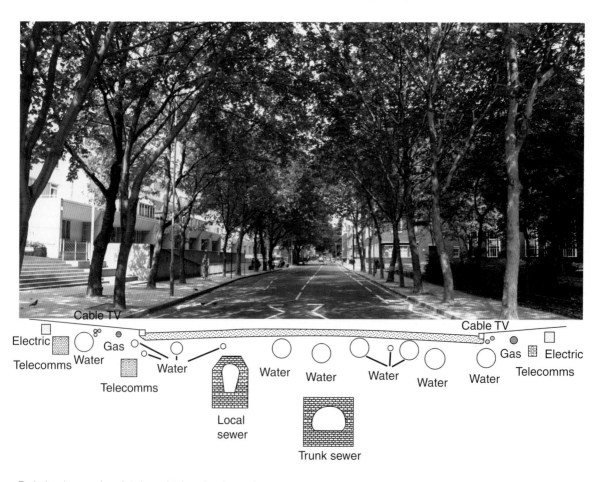

Typical underground services beneath a busy London road.

WATER USE BY TREE (ADAPTED FROM BRADSHAW *et al.*, 1995)

Crown projection (m^2) × LAI × mean evaporation × adjustment factor = litres per day

- Crown projection = horizontal surface area of branch spread (πr^2)
- LAI = Leaf Area Index; factoring in overlapping leaves to increase surface area (use 4 in UK)
- Mean evaporation = standard rate of water loss from a pan of water (approximately 3 in UK)
- Adjustment factor = because rate of water loss from leaves is less than from an open pan (0.2 in UK)

SOIL VOLUME NEEDED TO SUPPLY THIS DAILY WATER THROUGHOUT A 20-DAY DROUGHT (HIGH END OF UK CONDITIONS) (ADAPTED FROM BRADSHAW *et al.*, 1995)

(Daily water use) × (no. of days drought) × (available water holding capacity of soil) = minimum volume of soil required to allow the tree to reach the planned, ultimate size (if water use is in litres the soil volume will be in litres)

---

bend, snap or damage, but they are obvious targets and cost more to plant, protect and maintain. Vandalism to a tree may be due to it being on a route from a pub, being close to where a group of people hang out, or it may be in a place that is not appreciated by a local resident. Understanding these factors helps in establishing an appropriate response.

Dealing with the conditions above ground is only half the battle. Soil volume and quality are a major issue when considering street tree planting. Soil pH is likely to be very high and the soil may be a mixture of brick dust, cement and other pollutants. For the trees to thrive, it is vital that they are planted into a medium that can provide decent nutrition and with a volume sufficient to allow them to grow to their planned, mature size. Beneath streets, the ground is criss-crossed with drains and underground services, so finding space to pack in good-quality topsoil is not easy. But there are ways of providing a growing medium. One approach is to use tree soil; that is, soil that has been produced by mixing a careful set of ingredients that maintain pore spaces and provide nutrition even after compaction for hard surface installation. European tree soils tend to be based on using similar-sized sand particles. In the USA, tree soil is produced from crushed rock particles.

A generalized calculation is possible for working out the minimum soil volume to provide for the whole life of an urban tree. First, calculate the expected water loss of the tree at its ultimate, planned size (use the ultimate crown projection, which is the area of the crown in a horizontal plane), then assume a summer drought period of twenty days (which is at the high end of the spectrum of possibilities in Britain). Assume that the available water-holding capacity of the soil is 20 per cent (an average UK figure), then determine the minimum soil volume needed to provide water to the tree during such a drought. Bradshaw (*see* Bradshaw *et al.*, 1995) uses generalized values for leaf area index (LAI), mean evaporation and an adjustment factor for the difference between water evaporating from an open pan and water transpired through the surface of a leaf. In UK these three factors, multiplied together give a value of 2.4 (4 × 3 × 0.2). This equates to 2.4ltr per m^2 of crown. To provide this medium, the ground beneath the street is excavated and the tree-soil introduced and packed in place. The trees are then planted in their planting spots and a hard surface added.

The alternative approach is to use heavy duty plastic crates that can withstand the expected loads of the street or road surface. These are placed in the excavated areas beneath

SOIL VOLUME CALCULATION TABLE (ADAPTED FROM BRADSHAW *et al.*, 1995)

		Average UK conditions	Assume 20 days represents the maximum drought experienced			
			Assume available water holding capacity of soil (AWHC) is 20%			
	$CP = \pi \times r^2$	LAI x mean evap x adjustment				
Ultimate crown radius	Crown projection	4 × 3 × 0.2 litres /m²/day	Water use per day	Litres used over 20 days	Litres of soil @ 0.2 AWHC	m³ of soil
2.5	20	2.4	48	960	4,800	4.8
3	28	2.4	67	1,340	6,700	6.7
4	50	2.4	120	2,400	12,000	12
5	79	2.4	190	3,800	19,000	19
6	113	2.4	271	5,420	27,100	27.1
7	154	2.4	370	7,400	37,000	37
8	201	2.4	482	9,640	48,200	48.2

1,000 litres per cubic metre

the street and filled with good-quality topsoil. Above the crates, the footpath or road surface is constructed as normal. With both of these methods, it is important to include irrigation and aeration into the specification, so that roots growing through the soil have access to these resources.

Urban areas are characterized by hard surfaces, which means that most water runs off into drains, but some will also seep into the ground around trees. This is generally a good thing, but in some situations the flow of water can erode soil beneath the surface or water tables can become raised. Because of this, it is important to consider drainage as well as irrigation when planning street tree planting.

Trees and Subsidence

Tree growth requires water, which is drawn up into the plant body primarily from the roots. On

Root Cells into which topsoil can be loaded. The open areas are where trees will be planted. (Photo courtesy of Greenleaf)

Silva Cells showing large area for addition of topsoil.

clay subsoils the removal of water by tree roots can affect the volume of the soil. This is because clay swells as it absorbs water and shrinks as water is removed from it. In some cases, the extraction of water by tree roots is an important factor in clay shrinkage and subsequent damage to buildings. However, there are other causes of subsidence (and also other defects resulting in similar damage), so careful investigation is needed before concluding that trees are a major factor (Institution of Structural Engineers, 2000).

The London Tree Officers Association has done a lot of work to develop a realistic procedure for understanding the issues relating to tree-related subsidence damage to buildings. Their *Risk Limitation Strategy for Tree Root Claims* (LTOA, 2008) sets out the problem, provides local authority tree officers with a way of dealing with calls to remove trees and encourages policy-makers to provide resources to deal with this issue in a professional way.

While drainage is essential, these drains are likely to remove all water from around the tree.

Street Damage

Trees affect the surrounding infrastructure as they grow. In 1988 in Manchester 30 per cent of street trees were causing some damage to pavements and 13 per cent were damaging kerbs. In Hanover, Germany, in 2002 hardscape damage was found to be associated with about half the 2,881 street trees. In California $70 million is spent each year on repairing street tree damage to paths, kerbs and gutters (Costello and Jones, 2003).

Tree-management strategies should include an assessment of this problem and identify cost-effective ways of minimizing the damage. Management strategies begin with choosing the right tree to reduce future problems, but some remedial work to existing trees will inevitably be needed. There are ways of managing this issue that focus on the infrastructure rather than the tree. Pavements and kerbs can be designed to minimize conflicts with tree roots, particular materials can be used that do this and tree roots can be diverted from vulnerable structures and encouraged to grow in less critical areas.

Tree-Based Strategies

Tree-based strategies focus on either choosing the right species that will minimize future problems, or root pruning once damage has happened. Trees that have prominent buttress roots or are much wider at their base than further up the stem will cause more problems than those without these characteristics.

Existing trees have to be managed, however, and where root pruning is unavoidable it should be carried out carefully to minimize tears, rips and unnecessary wounds. A straight cut through the root will leave the smallest cross-section of wood exposed and limit opportunities for infection. Shaving roots, retaining only the bottom half, to give a greater clearance for infrastructure leaves a large cross-section of wood exposed and is likely to lead to infection and pathogens spreading into the base of the trunk.

Trees already showing signs of decay or stress, or those leaning, should only be root-pruned as a last resort. Root pruning broad-leaves when in full leaf will usually lead to branch die-back. Species sensitive to root pruning are likely to react badly at whatever time of year they are cut. Root pruning of large roots, or more than one large root, or pruning on more than one side of the tree, increases the risk of instability and such trees will need to be carefully monitored to assess the effects of this work.

Pop-outs help keep structure and traffic away from the base and major roots of trees.

Here trees have sufficient space (for the present) from adjacent properties, the road and the footpath.

Infrastructure Design

There are many ways to give trees more room for root growth. Not every one will fit each situation, but the objective of each is to maximize the distance between trees and infrastructure. At the tree planting stage, it is vital to ensure that the exploitable soil volume is sufficient for the ultimate size of the tree chosen. Above ground, the same consideration is needed. The open ground around the tree must be planned to be sufficient for the whole life of the tree in order to minimize the potential for damage from trunk expansion, buttress development, or surface rooting. For trees of large ultimate size, a width of 5m (16ft) is recommended, but planting strips should be at least 3m (10ft) wide. For individual trees, the minimum opening recommended is 2 × 2m (6.5 × 6.5ft) to minimize pavement damage (see Costello and Jones, 2003).

Pavements (sidewalks) can be curved away from or around trees and paving slabs can be of smaller size to avoid installing hard elements too close to the tree. In some cases, extra space can be excised from the road to form 'pop-outs'. This is particularly suitable as part of a traffic-calming scheme, or where the road edge is zoned off for parking. The principle of pop-outs can be extended to groups of trees forming tree islands where roots from all trees exploit a common soil volume. In some streets the pavement may need to be eliminated to provide this space. Planting on one side of a road may be all that is achievable if trees are planned to thrive and be properly inserted into the street scene. Where important tree roots are encountered during resurfacing or repairs, the new path or drive can bridge or ramp over them using piles and beams to stabilize slabs. Reinforced concrete slabs can be formed on these piles, provided sufficient soil or gravel material isolates the roots from the concrete.

Retaining voids between the surface and the roots may be unacceptable due to the accumulation of litter or the risk of encouraging vermin, but, where necessary, a void can be filled with large-size gravel that retains a large volume of pore spaces. Any piles, obviously, need to be installed without damaging tree roots.

Where the whole new surface is proposed at a higher level than the natural ground in which there are trees, the bridging principle can be extended and the surface can be suspended like decking above the ground to prevent soil compaction. Alternatively, the area around the tree can be designed so that the infrastructure stays back from the tree stem. In most cases, a design like this will need to incorporate grilles or grates so as to cover safely the space between the tree stem and the surrounding surfacing.

Modern materials and installation practices can help in designing space for trees. Permeable concrete is now available, different reinforcement materials can be used to add strength to paths and drives to resist root growth and expansion joints can be used to allow differential lifting of slabs in order to limit damage. Tarmac and asphalt are more flexible than concrete, but also weaker. Permeable tarmac, which allows water and air through the surface, can be specified, although it is likely to clog up with soil and debris over time.

Paviours have the advantage that they can localize tree-root disruption and can be easily lifted and replaced. However, it does mean that they quickly become trip hazards when displaced. Permeable pavers are now available with nubs around the edges that increase spaces between. However, permeability can still gradually decrease.

Root Zone Strategies

Products are available that direct root growth away from sensitive areas. These root barriers can be rigid or flexible and used around the edge of a planting pit or along the side of a service trench, for instance. This approach can only be successful for the trees if the roots are directed to soil that is of sufficient fertility and volume to sustain growth. Continuous trenches are like tree islands underground. Where a line of trees is planted, a continuous trench filled with structural tree soil or topsoil packed into open plastic crates reduces their vulnerability to drought, or growth limitations.

The open plastic crates can be used to connect areas of open ground with rootable soil beneath a pavement or drive, allowing tree roots to grow from an area of limited soil volume into additional uncompacted soil where they have access to further resources.

In highly urbanized areas such as city centres, existing trees that are under pressure because of the scarcity of exploitable soil can be helped by lifting sections of the hard surfacing to find and expose tree roots. The compacted ground around the roots is removed using pressurized air and replaced by good-quality topsoil. The hard surface then needs to be replaced without compacting the soil. If the soil cannot be adequately compacted to support the surface, structural soil can be used around the roots, although this will not be as beneficial to the tree. Root paths are an extension to this idea. Trenches 10cm × 30cm (4 × 12in) are installed in compacted ground around trees and these are filled with good-quality topsoil and a vertical drain. These trenches become routes of preferential root growth due to the less compacted soil and greater aeration.

Pressures on Young Trees

Below-ground management is not usually programmed by the tree manager and may well not be under his or her control at all. Root pruning of urban trees for arboricultural purposes is not a common practice in the UK. Far more usual is the repeated digging up of roads and footpaths by statutory undertakers to make repairs and to install new services. Often the tree manager is not informed of such programmes of work; in fact, it is unusual if this coordination does take place.

Strategies for limiting road salt damage need to include good liaison and communication with the highways authority, consideration of alternative gritting materials, changes to the way salt is stored and applied, amelioration works to limit damaging effects and species choice for new plantings.

PARKS AND WOODS

Away from roads, most public areas containing trees will be parks or woodlands. Parks are usually a mix of vegetation types, water and hard features, but where there are trees these usually form important, if not dominant, features. The

Irregular trees and groups add to informality of an area.

main objectives for park areas include low maintenance costs and ease of use and safety for visitors. Trees in parks may be individual specimens, boundary groups and lines, or closed-canopy groups in less-frequented parts of the area.

Woods are often seen as being a part of parkland, but different in that they cover a larger area, with less intensive use that can be managed in a distinctive way.

Trees in urban parks add interest because of their three-dimensional character. This is particularly valuable in built-up areas as it helps to block and screen the surrounding buildings, giving a feeling of a more natural, relaxing area. The ability of trees to represent the natural world is heightened in urban surroundings and the way that they alter over the year helps to give city dwellers a reference point for the changing of the seasons.

The open spaces of a park provide opportunities for planting a wider range of tree species than is possible along city streets. There are places where specimen trees can be allowed to grow to their natural size and grandeur and trees with particular features, such as blossom, autumn colour or showy fruit can be planted and displayed to maximum effect by controlling the adjacent vegetation.

However, too much species variation can detract from the character of an area. Even in an arboretum, the trees are usually grouped according to their botanical family or particular characteristic. It is usually aesthetically pleasing to have a limited number of tree species in any one area. This increases a sense of character for that area and helps to make it distinct from another place. So, if a wide range of species is needed or desired in a park they are more likely to have a positive impact on the park users if

A veteran ash. Not a perfect specimen, but an interesting tree, an ecological feature and a link with the past.

they are grouped to give distinct and different characters to the various parts of the park in which they are planted.

Parks

Individual trees in parks are often surrounded by grass. They may be rooted into very compacted ground, due to hundreds of feet pounding over it daily, or restricted in root growth because of nearby structures. In many cases, the trees were there before the park!

Trees in Park Design

Trees in parks are found as individual specimens, in groups or clumps of various sizes, in avenues or rows, or as shelterbelts. Single trees can act as focal points to attract people, or to attract the eye. Particular species of beauty can be used, or a weather-beaten specimen may be full of character. Tree groups provide large blocks of foliage and block or frame views. The size does not mean the space beneath is unavailable. Pruning or management of the understorey can create shaded areas still retaining views. Clumps tend to be informal and contribute to a natural, informal landscape. Greater irregularity accentuates informality, while regular spacing, uniformity of size or species adds to the formality of an area.

Avenues are the most formal example of tree groups and are used as an architectural feature; complementing buildings, or creating an architectural effect where there is none. Lines and rows of trees tend to focus views and join up other elements in the landscape, or define a space.

Successful shelterbelts include a range of size of trees and shrubs and so fill the space from ground level to the treetops. They provide shade from the sun, shelter from the wind and help to absorb noise.

Crown-Management Options

Trees growing in parks will only infrequently be in locations where their crowns are able to grow unrestricted to their natural, ultimate size. More likely, the trees will either be growing close to roads or paths, or be close to buildings. Where trees and people are in close proximity, the trees usually have to give way.

The crowns of park trees can be managed by removing the lower branches to allow free access for people and vehicles beneath. This 'crown lifting' is a routine tree surgery operation that does not seriously harm the tree if the diameter of the wounds is kept small. A successful crown-lifting operation should remove a maximum of 15 per cent of the existing live crown and, ideally, the remaining live crown should make up two-thirds of the tree height. It may be appropriate to treat both sides of the tree in the same manner, but there is no biological or structural need to cut branches on the opposite side from the road. The tree will have to adapt to the changed weight and wind-sail effects, but should reach a dynamic balance over a couple of years.

The appropriate clearances above ground level depend on the use of the road or path. Pedestrian, cycle and equestrian paths have the same requirements whether inside or outside a park. Roads may be designed for only low vehicles or for buses and lorries. In a park, crown lifting to a height of, say, 4m (13ft) may be reasonable because of the type of vehicle using the road. However, many roads need to be accessible to fire engines and this may affect the crown-lifting specification.

Trees may also obscure a view as they grow. In such an instance it may be reasonable to reduce the height of the trees in order to prevent the view being obscured. Crown reduction may result in new branches that are less well anchored to the main branch structure, so this operation should only be carried out where it is clear that it is the best way of achieving the management objectives. However, when such pruning is necessary, it is best implemented as a pruning regime as early in the life of the trees as possible, so that the size of wounds can be kept to a minimum.

Where trees are damaged by climatic extremes, crown reduction may be necessary to help restore a safe, pleasing shape. This work may take both a number of operations and several years before an effective new crown is achieved. Patience is needed, along with periodic reassessment and arboricultural skill.

Branches can be kept away from buildings by crown pruning. In many local authorities a separation of between 1–2m (3–6ft) is considered reasonable. As branches extend each year, a crown-pruning operation is likely to be needed periodically; it is not usually a one-off maintenance job.

An electricity substation is a specific type of structure and one that is relatively common in and around parks. The pruning of trees close to it will require the electricity provider's guidance on safe working to be followed. Low branches may overhang and sway onto the equipment and dead branches may drop into the substation. For these reasons it is necessary to keep nearby vegetation under control.

In parks there are often many different structures. Where trees and structures are in close proximity, either or both can be damaged. British Standard BS 5837 includes guidance for safe distances for planting trees close to buildings or walls, but it does not say anything about existing trees. However, the knowledge used to set the distances can also help to determine when a tree is too close to a building. Provided there is a 2m (6.5ft) gap between a tree and a wall, most damage to the structure by root and trunk growth will be avoided. In some situations, this minimum separation distance can be reduced to 1.5m (5ft) (BSI, 2005) (*see* table on page 136).

Woodland Management

Overview

Woodland management consists of assessing the stage of growth of an existing wood and making appropriate interventions over time to achieve the set objectives for that area. If the end point of the existing trees is for them to be harvested for timber and replanted, this will guide management and maintenance operations. Where recreation or protection is the main objective of the land use, the continuation of the existing mix of species and tree size and age is likely to be appropriate and this will affect what operations are carried out.

A general principle followed by foresters is that as trees grow, the number on a set area of land must decrease if those remaining are to be well formed and healthy. Trees growing close together will affect each other as they get bigger. Branches will grow into the same space and roots will exploit the same soil. Periodically, the woodland manager has to assess the trees and remove some to make space for those remaining.

In recreational woods it may be that small groups of trees can be left with crowns that form a single, visual feature, but the natural course of growth will favour the strongest-growing trees and these will gradually suppress the weaker ones until they die. Where the woodland manager needs to act is when the suppressed trees become a hazard to users of the wood, or where potentially good-quality trees are affecting each other so that all of them are likely to become defective and vulnerable to future damage.

Where there are roads or paths through woods these are opportunities for increasing wildlife habitats, but they also increase the need for work on the trees so that they do not become unsafe. Large lumps of deadwood hanging over well-frequented paths or roads will need to be periodically pruned out and the usual statutory heights must be maintained to allow people and vehicle access.

Management for wildlife focuses on light and on appropriate plant species. Native species harbour and encourage much greater diversity of insects, small animals and other plants than do introduced species. And since a greater range of plants can be grown where light levels are high, the edges of woods, and especially along roads and paths, are prime areas for encouraging shrubs and ground flora and their associated ecology. By scalloping the edges

Too many trees on a small space. They need to be reduced in number to allow the better specimens to thrive.

A path through an urban wood.

of paths and roads, a greater area is made available for wildlife. Where the road or path is orientated north–south, the width of the open corridor needs to be greater than when it is orientated east–west. This is because a north–south road will suffer from shading trees to the east and west. An east–west path will benefit from receiving sun for a larger part of the day.

Planting new woodlands can also use this principle to encourage wildlife by careful orientation of paths and leaving open areas within the woodland itself. The numbers of trees planted, however, should be kept relatively high, as a primary object for planting a woodland is to 'capture the site' as quickly as possible. By planting a high number of trees close together (2m [6.5ft] between trees is a usual spacing), the young plants can shade out weeds and limit competition for water and nutrients. In this way, the site is 'captured' for tree growth.

Once the trees have become established and they are growing strongly, they will need to be thinned out, so that those remaining have sufficient space to continue growing. The higher the density of trees planted, the sooner will be the need to thin them out. This continues throughout the life of the trees and we are back to the principle of reducing the numbers of trees per set area as they grow larger.

Woodland Design

New woodlands are likely to be more successful if they reflect and use the natural landform. For instance, if woodland covers hills and ridges, while valleys and hollows are left open, this will emphasize the undulations and natural folds of the land and increase the microclimate benefits of the sheltered areas. Alternatively, woodland in valleys creates more shelter and shade, leaving open areas near hilltops with wide views into the distance. So studying the area before planning and planting is vital.

All woods need open areas as well as densely planted ones. Variation in the mix of trees and open space is usually a positive thing. Open areas can be concentrated near to entrances and places where people congregate; the areas further from structures and people can be planted more densely and left wilder, giving the feeling to visitors that they are leaving the city further and further behind as they explore the woodland.

The transition from woodland to the land use of the adjacent areas should be carefully considered. Housing areas may be adversely shaded by dense woodland close to them and woodlands abutting fields should be blended into the new landscape by using hedges and smaller growing species.

Woodlands are not all the same. For instance, some woods contain mostly tall, mature trees, while others are regularly coppiced, or are composed of widely spread trees with pasture or bushes between. Each type has its own character and requires particular management and maintenance. Different species form differing types of woodland, too. Pioneer trees, such as birch, create open canopies, letting a fair amount of light through, whereas beech shuts out most light and limits what can grow beneath it. Most natural woodlands in the UK are formed from one or two predominant species, with others putting in occasional appearances or encountered in groups where conditions change, such as less or more fertile soil, higher water table, greater exposure and so on. Planting can create these variations, with planning taking its cue from the details of the site and the objectives of management. Natural colonization from seeds in the soil cannot be relied upon in urban areas, but, where it occurs, it can be a valuable addition to plantings, adding maturity or naturalness to an area.

Open spaces within a woodland are important as they increase the variety and diversity of the area and provide ways of utilizing land where trees will not, or must not, grow, such as beneath pylons, above gas pipelines, or where contamination is severe. Roads and paths can be used to draw people into the woodland and to control their use of it. The numbers of visitors and the size of the area obviously affect the intensity of use; also, the ways in which

---

FENCING HEIGHTS

- Sheep can be kept out of an area of tree planting using standard stock fence which is 90cm high. This will need a spring-steel wire above this height to bring the fence to a height of around 1m.
- Rabbit fencing is wire netting 90cm in height but the bottom 15cm is turned outwards to prevent them burrowing underneath.
- For small deer (roe and muntjac) rabbit netting is used but above this, to a height of 1.8m, added spring-steel wires or netting is used.
- Red, fallow and sika deer need a 2m tall fence with more spring-steel wires for added strength.
- Posts and rails should be pressure-treated.

---

people use the site also affect the design of access routes. Small woods may need multi-use paths, but larger woods may give opportunity to separate walkers, cyclists, horses and other users. Where personal security is a real or perceived issue, wider, straighter paths and lighting may be appropriate.

Restoration of existing woodlands deals with the same issue as above. Paths, roads and edges may be regular in pattern, in poor condition, or not suited to a new management vision. Open spaces may be needed in different places and more trees needed in worn-out, neglected tree stands or to increase diversity.

Tree Protection

Individual tree protection can take the form of tree shelters, or mesh or spiral sheaths. Spiral sheaths can reduce bark stripping by rabbits. Plastic mesh guards can be of a range of diameters. These polyethylene netting tubes provide protection against rabbits, deer and livestock, but not voles.

Tree shelters are translucent polypropylene tubes, usually 1.2m (4ft) tall, that provide protection for newly planted trees from wind and from grazing pests such as rabbits and deer. But they are expensive and when used they are often funded by reducing the numbers of trees planted. This increases the spacing between trees and lengthens the time taken for site capture. The shelters may attract unwelcome attention, in the form of vandalism or letters of complaint about the disfigurement of the landscape, but they can be cost-effective.

The alternative to tree shelters is fencing, but this is not a cheap option. Where fencing is used to protect trees, it is most cost-effective where the area is in the shape of a square, maximizing the inside area per length of fence. So one way of reducing costs is only to fence rectangular areas and to use tree shelters outside the fence, where irregular edges would require wiggly, expensive lengths of fence.

Fences and tree shelters are temporary measures. Once the trees have become established they are less vulnerable to damage from grazing animals. Typically, this is after the trees have formed their crown above the tree shelter, or when the crowns are beyond the reach of deer. Roe deer reach up to around 1.2m (4ft), but there are many sika and hybrid deer near to urban areas and these can reach up to 1.5m (5ft). Where snow lies thick in the winter, this can mean that higher tree branches are at risk from these animals.

Animal Pests

When their populations are high, voles and field mice can cause considerable damage to young trees by stripping the bark from the roots and lower stem, up to the height of the surrounding vegetation. Damage can occur at any time of the year, but is most likely in late winter.

Rabbits strip bark and will damage buds and shoots up to a height of around 50cm (20in). Most of this damage occurs in the winter and early spring. When snow lies on the ground for a prolonged period, the rabbits can reach up higher. Hares roam widely and, when allowed to become numerous, will damage young trees by biting off the young shoots.

Badgers are beneficial, as they eat young rabbits, mice and voles. The damage they cause to trees is mostly because they dig under fences or break them down, allowing rodents in to munch on the young trees. Where this happens, the pragmatic response is to build a wooden badger gate in the fence that allows the badger through, but is too heavy for the smaller animals to shift.

Grey squirrels are rodents that climb trees. They are therefore able to damage trees at the base and also in the crown. The vulnerable trees are those with thin bark from the ages of around ten to forty years. The damage at the tree base is similar to rabbit damage, but occurs usually from May to July. Beech, birch, ash, sycamore and Norway maple are most susceptible, although oak and sweet chestnut can also be affected. Corsican and Scots pines are the conifers most at risk.

There are six kinds of deer in Britain at present: roe, muntjac, fallow, sika, Chinese water deer and red deer, but some hybridizing is going on, too. In urban areas, you are most likely to encounter the roe or sika deer. Deer damage trees by browsing on the leaves and shoots. Roe can reach to 1.2m (4ft), fallow and sika 1.4m (4.5ft) and red 2m (6.5ft). Muntjac reach only to 50cm (20in), but they also bend whippy stems to reach higher shoots. Trees can

also be damaged by fraying, when males rub off the velvet skin covering their antlers or mark their territory. Red, fallow and sika also damage trees by bark stripping, pulling the bark up the stem from the base.

Shelterbelts

Planting trees in a long, thin strip of land can provide shelter against the wind and a barrier against noise. However, the benefits are dependent upon many factors. Noise barriers can be psychologically effective even though the reduction in decibels is quite modest (5–10 decibels). Species, thickness of the belt, topography and wind direction are also important factors.

Shelterbelts are most effective when they are composed of a mix of conifer and broad-leaf species and let around half of the wind go straight through. They have been found to be effective for a distance of twenty to thirty times their height on the lee side and they even provide some protection, three to five times tree height, on the windward side. The greatest benefit is felt five to seven times tree height downwind of the shelterbelt. So, a shelterbelt of trees of height 20m (66ft) will be most effective at a distance of 100–140m (330–460ft) from it. Obviously, such a shelterbelt needs to form a barrier to the wind over its whole height, so it is best to have a mixture of species, including tall trees, small ones and tough shrubs.

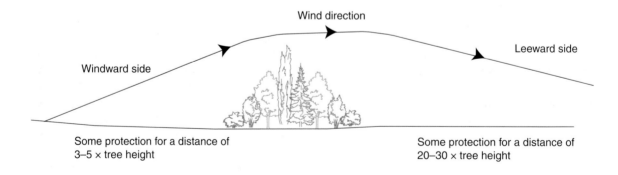

Some protection for a distance of 3–5 × tree height

Some protection for a distance of 20–30 × tree height

A shelterbelt/treebelt design. (Adapted from Starr, 2005)

Shelterbelt/treebelt with trees in the centre growing to 10m tall and shrubs on either side growing to 2–3m tall

Distance between shelterbelt/treebelt and the road should equal the treebelt width

The length of the shelterbelt/treebelt should equal or exceed double the distance between the houses and the road

To reduce moderate traffic noise in communities use a shelterbelt/treebelt of 7–17m width

To reduce heavy traffic noise in suburban areas use a shelterbelt/treebelt of 20–35m width

Road or noise source

Length A

Length A

Length A

Length A

Houses or other places requiring noise reduction

LEFT: Treebelts for reduction of traffic noise in communities and suburban areas. (Adapted from Cook and Haverbeke, 1974)

A wide strip of shelterbelt also allows some leeway when planning its replacement. Half the width can be removed, with the cover provided by the remaining half helping to protect the newly planted trees and shrubs. If space is not limiting, a whole new shelterbelt can be established on the lee side of an existing belt before it is removed.

OTHER PUBLIC SPACES

Operational Land

Overhead Power Lines

Urban areas depend on power supplies and telecommunications. These are most obvious in the landscape in the form of electricity pylons and telephone cables. As trees grow, they can interfere with these items of infrastructure and so they need to be periodically cut back to prevent damage.

Planning law allows a statutory undertaker to carry out work to safeguard the continued provision of its supply. The Electricity Act 1989 places a duty on electricity suppliers to keep trees clear of overhead power lines for the safety of the public. District Network Operators (DNOs) are also required to keep trees from interfering with the electricity supply and to improve reliability by maintaining a sufficient distance between trees and overhead power lines. The Telecommunications Act 1984 places a similar duty of care on telecoms providers.

Specifications for maintaining clearances between trees and pylons, or telegraph poles carrying electricity, vary according to the voltage level of the power line. Branches able to support a person's weight or a ladder should be kept at a minimum 3m (10ft) distance from

The wider a shelterbelt is, the better. A strip of 30m (100ft) width is very effective and allows a range of species and sizes to be planted, low-growing at the edges to tallest in the centre. Such a shelterbelt will be robust and effective over a long period. An irregular shape of belt and species mix will develop a natural character, but plants must be kept growing well, as gaps will weaken its beneficial effects. The ideal length of shelterbelts is ten to twelve times the width. Short belts are less effective because the wind tends to swerve around them. So, a 30m (100ft) wide shelterbelt would ideally be around 350m (1,150ft) long. Where the shelterbelt is planted for noise reduction, Konijnendijk et al. (2005) suggest the length should be twice the distance from the noise source to the noise receiver.

11kV or 33kV overhead electricity cables near trees.

Overhead telephone cables.

voltages up to 33kV, and smaller branches should be kept at least 0.8m (32in) distant. As the voltage increases to 110kV, 275kV and 400kV, the minimum clearance distances increase. How far vegetation is cut back during tree-pruning operations depends on the expected return period. Any pruning must be designed to keep the power lines safe until the next visit, but responsible management will also minimize the work needed to reduce immediate costs and to comply with the requirement to preserve the amenities of the area. ENA TS 43-8 is the technical specification containing these standards, but the DNOs also have a statutory obligation to maintain these minimum clearances and to ensure there are no tree-related power cuts.

Work to trees and shrubs near to electricity cables must be carried out in accordance with BS 3998:2010, *Tree Work – Recommendations,*

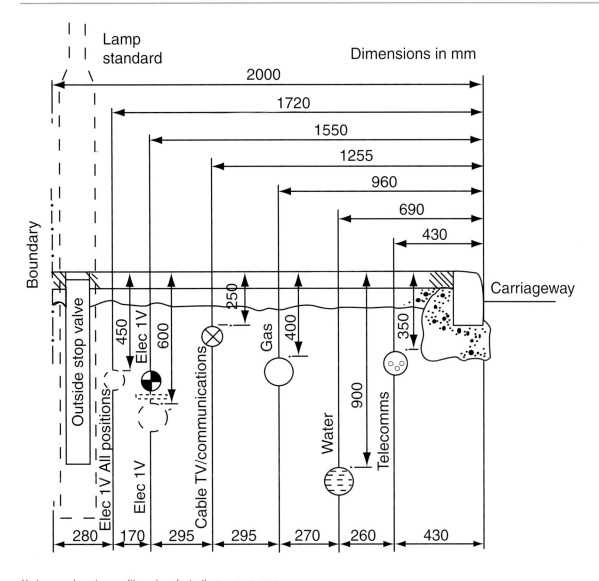

Underground services profile under a footpath. (From NJUG, 1997)

and the Electricity Act 1989 requires the amenity and beauty of the landscape to be protected.

There are no clear minimum distances for maintaining trees near to telephone lines. The Telecommunications Act 1984 merely states that work should be done 'in a husbandlike manner' and should cause the minimum damage to the tree. This Act applies only where trees overhang streets and gives the statutory

undertaker power to remove obstructing branches, or to require the tree owner to do so.

Underground Services

For underground services, the National Joint Utilities Group (NJUG) provides guidelines for working close to trees to install or repair apparatus (*NJUG Volume 4*). The area around any tree close to excavation is classified into one of three zones: prohibited, precautionary and permitted.

The prohibited zone is within 1m (3.3ft) of the trunk of the tree. No excavations should be done here and the area should remain clear of materials, machines and equipment and no vehicles should pass over it. Where excavation is unavoidable, roots should be protected by dry sacking and the trench backfilled with granular materials as soon as possible on completion of the work, with the sacking removed before backfilling.

The precautionary zone is defined as being within a radius of the tree equal to four times the circumference (measured at 1.5m [5ft] above ground level). Excavations within this zone should be by hand, or using trenchless techniques, such as a mole or thrust borer. Roots over 25mm (1in) in diameter should be retained, unless advice has been provided by a qualified arboriculturalist or the local tree officer. Heavy machinery should be kept off any open ground and materials, spoil and chemicals should be kept away.

The permitted zone is outside the precautionary zone. Here, machinery can be used for the excavation, but care is still needed in case roots are encountered. Roots over 25mm (1in) in diameter that need pruning should only be treated after consultation with a qualified arboriculturalist or the local tree officer. All the precautions mentioned above should also be considered.

Any roots that are to be pruned in any zone should be cut using sharp secateurs or a handsaw, in accordance with best arboricultural practice. Backfill on highways sites should be with an inert, granular material with topsoil mixed in. On non-highways sites, it is best to use the existing soil as a backfill material. Any exposed roots should be protected with dry sacking until backfilling.

Where extensive programmes of excavations for underground services are planned, some method is needed to avoid damage to vulnerable trees. A rapid survey, carried out by arboriculturalists, prior to finalization of the trenching routes, can classify into three categories all trees along proposed routes. High-risk trees are those that will be seriously affected by disturbance. Medium-risk trees are those that are likely to be moderately affected by the trenching. Trees in the low-risk category are considered unlikely to be affected by root disturbance. The trees are assessed using the following criteria:

Underground services can easily reduce tree root space and make life difficult for roots.

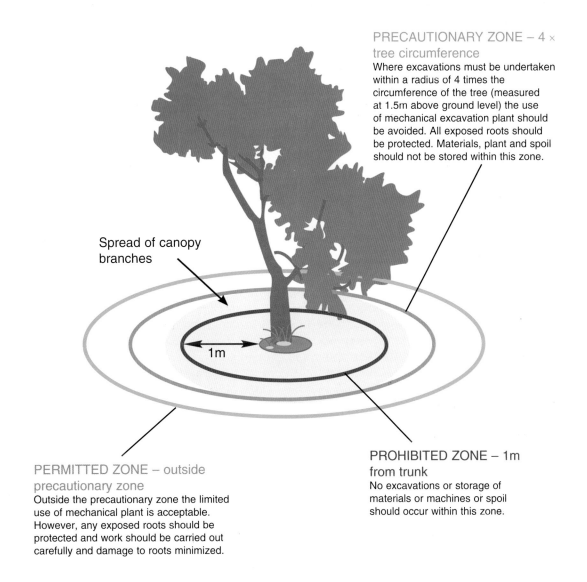

PRECAUTIONARY ZONE – 4 × tree circumference
Where excavations must be undertaken within a radius of 4 times the circumference of the tree (measured at 1.5m above ground level) the use of mechanical excavation plant should be avoided. All exposed roots should be protected. Materials, plant and spoil should not be stored within this zone.

Spread of canopy branches

1m

PERMITTED ZONE – outside precautionary zone
Outside the precautionary zone the limited use of mechanical plant is acceptable. However, any exposed roots should be protected and work should be carried out carefully and damage to roots minimized.

PROHIBITED ZONE – 1m from trunk
No excavations or storage of materials or machines or spoil should occur within this zone.

NJUG tree-protection zones.

- tree species, age, vigour, size and condition
- height of crown base
- past tree management
- adjacent surface cover (paved or soft)
- ground levels and water table
- exposure and slope
- other recent construction work

- number of ducts to be installed.

This assessment allows an appropriate course of action to be taken to avoid damaging trees, such as: re-routing trenches to avoid the trees completely; diverting trenches around the trees; or following appropriate precautions when trenching in vulnerable areas close to trees (see Browell, 1996).

Shopping Areas, Colleges and Hospitals

These land uses are grouped together because the issues of managing trees within them are the same in kind, but with a slight variation in emphasis. Shopping areas, college campuses and hospitals typically include substantial buildings at the centre of a complex surrounded by parking and access roads, with landscaping of the perimeter and open areas.

The benefits provided by trees in all of these areas are similar, but the psychological effects of natural vegetation are more pronounced when considering patients who remain in hospital for short or extended periods. For recovering patients, trees and natural features can have a real effect on their well-being, reducing recovery periods and improving mental health. On college campuses, the quality of the environment can have an effect on the well-being of the students, while in shopping areas the psychological benefits are still valid, as it is easier to encourage shoppers to an area that includes significant natural features where they are likely to feel less stressed and happier.

The opportunities for tree planting and the need for tree management are similar. Trees can: provide shading of car parks and buildings; modify ground temperatures; reduce flooding; provide shelter from wind; and screen unwelcome views or frame attractive ones. The issues to consider include the location of new trees, so that sight lines are not impeded, and to ensure that buildings, services and security measures such as CCTV cameras operate effectively.

Large areas of car parking provide an obvious opportunity for trees to add aesthetic value to an area, as well as providing other environmental benefits. But car parking plans often shy away from tree planting due to installation costs and reduced numbers of parking spaces. The case needs to be made more strongly that inclusion of trees is not a cost, but an investment in the future.

Where trees are planted around the windward sides of a parking area they provide shade and shelter. Here, the tree roots can utilize the open ground at the edge of the hard surface. A hard edge to the car park will be needed to prevent informal parking on the open ground and planting should be set back from the parking edge to avoid future damage from overhanging car bumpers. Existing trees should

Trees in a car park. No protection or space allocated for trunk growth and likely not enough space provided for roots either.
(Photo courtesy of Jo Ryan)

be given space around them to avoid this collision damage. To reduce damage during construction, a minimum root-protection area (BS 5837) should be protected.

Within a parking area, trees can be sited in islands that take up surface area and so reduce the number of achievable parking spaces. Alternatively, they can be slotted in between spaces, with protection bars or bollards installed to keep vehicles away. Strips of planting area can be designated where lines of trees are proposed.

As the above-ground surface area for the tree diminishes, it becomes more important to provide and prepare a below-ground rooting space. As ever, the ultimate, planned size of tree should dictate the soil volume needed. To minimize damage to kerbs, the minimum width of any tree island or planting strip should be 3m (10ft). Widths of 1.2m (4ft), which are common, have a very high damage potential, as well as making provision of adequate soil volume difficult.

Where there is no opportunity to remove whole parking spaces from hard surfacing to make space for trees, bollards and bars will be needed for tree protection from vehicles. In such cases, the trees should be placed between lines of parking spaces rather than between two spaces where car doors could do damage when opening. Soil volume is still the main factor to consider and, where the hard surfacing reaches close to the tree base, rootable soil must be installed beneath the new parking surface either using structural soil, or topsoil loaded into soil crates. Irrigation and aeration pipes should be provided, along with drainage at the bottom of the soil pit.

Porous surfacing is becoming more common and will become more important over time with increasing climate variability. Increased permeability will raise the moisture content of the soil and benefit tree roots, but drainage will still need to be considered to avoid waterlogging.

Open areas, banks and boundaries are obvious areas that may be planted with trees, or

protected because they contain existing trees. Space above ground should be considered so that trees are not located beneath power lines or other features, requiring maintenance or preventing adequate growth.

Pedestrian areas offer opportunities for retaining or planting specimen trees, but any investment in such work needs to ensure that everything (soil volume and quality, irrigation, aeration, drainage, protection, above-ground space and a suitable wearing surface) is provided to enable the tree to flourish.

Cemeteries and Churchyards

Cemeteries are oases of calm, greenness and diversity in urban areas. They often date from the beginning of the rapid growth of the town, or even earlier, and usually have a history of being managed without large-scale disturbance or chemical pesticides. All of these factors make them precious wildlife refuges and vital parts of the urban life-support system.

Cemeteries include particular structures, gravestones and memorials that are invested with much sentiment. On the one hand, tree

A beech engulfing a grave.

growth can cause damage to structures and may need to be controlled, but, on the other, the trees demonstrate the ability of nature to embrace our lives and be an agent of time's economy.

Although tree growth may be tolerated in some situations, possibly where an old gravestone is gradually being overwhelmed by a swelling tree base, there will be many instances where tree growth is in an inappropriate place and here trees will need to be removed. Maturing trees will accumulate deadwood, creating a hazard for visitors, and low branches may well obstruct the use of the area. Along boundaries, trees may cause problems for adjacent land. A fringe of unrestricted trees is a valuable landscape feature, but increasing use of land often leads to new houses being built with small gardens right up to cemetery borders. Here, some flexibility in managing the trees is needed. Any work to trees within such a site for these, or other, reasons will need to be carried out with extra sensitivity to avoid damage to memorials or creating nuisance to visitors.

Cemeteries are, in general, under the control and management of the local authority. Churchyards are under the control of each Parochial Church Council (PCC), whether the site is receiving new burials, or is closed. However, the PCC may decide to pass management of a closed churchyard to the local authority. Church of England regulations (Canon F13(2)) require all churches and chapels to be kept in an orderly and decent fashion, as becomes the House of God. This principle extends to the churchyard and the boundary of the site and is carried over into secular legislation.

All maintenance work to trees in a churchyard that is under the management of the PCC (except very minor or excluded works) must be presented to the archdeacon or consistory court, who will then issue a 'faculty' for the approved works. This procedure does not over-ride the need to apply to the local planning authority if the trees are protected by a tree preservation order, or if the site is within a conservation area.

Canals, Rivers and Lakes

Water features within urban areas are valuable as contrasts with the surrounding hard infrastructure. Where vegetation is included, the aesthetic benefits are greatly increased. From a tree-management perspective, canals, rivers and lakes provide opportunities for using trees in architectural ways, framing views, creating reflections in water and interposing between the water and buildings. But there are also difficulties in managing trees near water.

Tree roots can be disruptive of embankments and trees may grow too close to the edge of a watercourse and become a hazard. However, large-scale embankments and concrete retaining structures are unlikely to be disrupted by roots, if trees are provided with 1–2m (3–6ft) of separation.

Where work is proposed to trees over water it is always prudent to operate with at least two workers. This is not a situation to recommend lone working.

REFERENCES

Bradshaw A., Hunt B., Walmsley T., *Trees in the Urban Landscape* (E. & F.N. Spon, 1995).

Browell M.F., 'Tree Risk Assessment', *Arboriculture Journal*, 20, 3–12 (Arboricultural Association, 1996).

BSI, *BS 5837 Trees in Relation to Construction* (BSI, 2005).

Church of England, *Of the Care and Repair of Churches*, The Canons of the Church of England, Canon F13(2), Sixth Edition, including second supplement (Church House, 2008).

Cook, D.I., Van Haverbeke, D.F., 'Tree-Covered Land-Forms for Noise Control', *Research Bulletin*, 263. Rocky Mtn Forest and Range Experiment Station, USDA Forest Service, University of Nebraska, Lincoln, NE, USA, presented in *Trees as Noise Buffers*, The Overstorey eJournal No.60 (Permanent Agriculture Resources, 1999–2009, http://www.agroforestry.net/overstory/overstory60.html, 1974).

Costello L.R., Jones K.S., *Reducing Infrastructure Damage by Tree Roots* (Western Chapter ISA, 2003).

Crane B., *Avenues and Arboricultural Features*, Seminar XVI (Treework Environmental Practice, 2010).

DoT, *Well-Maintained Highways* (TSO, 2005).

Energy Networks Association, ETR 136, *Vegetation Management Near Electrical Equipment – Principles of Good Practice* (ENA, 2007).

Energy Networks Association, ENA-TS 43-8 'Overhead Line Clearances' [Issue 3, 2004] (ENA, 2008).

Energy Networks Association, G55/1 and G55/2 *Safe Tree Working in Proximity to Overhead Electric Lines* (ENA, 2000 and 2008).

Hart C., *Practical Forestry*, 3rd edn (Alan Sutton, 1991).

Hibberd B.G. (Ed.), *Urban Forestry Practice* (HMSO, 1989).

Highways Agency, *Design Manual for Roads and Bridges* (*DMRB*) (The Highways Agency, undated).

Highways Agency, *Landscape Management Handbook*, HA 108/04, Volume 10, section 3 of *DMRB* (The Highways Agency, 2004).

Highways Agency, *Road Geometry*, *DMRB*, TD27/05 (The Highways Agency, 2004).

Highways Agency, *The Geometric Design of Pedestrian, Cycle and Equestrian Routes*, *DMRB*, TA90/05 (The Highways Agency, 2005).

Highways Agency, *Geometry of Roundabouts*, *DMRB*, TD16/07 (The Highways Agency, 2007).

Highways Agency, *Traffic Signs Manual*, Chapter 8 (The Highways Agency, 2009) (http://www.dft.gov.uk/pgr/roads/tss/tsmanual/tsmchap8part1.pdf).

HSE, AFAG 804 *Electricity at Work: Forestry and Arboriculture* (HSE, 2007).

Institution of Structural Engineers, *Subsidence of Low Rise Buildings*, 2nd edn (ISE, 2000).

Konijnendijk C.C., Nilsson K., Randrup T.B., Schipperijn J., *Urban Forests and Trees* (Springer, 2005).

LTOA, *A Risk Limitation Strategy for Tree Root Claims* (LTOA, 2008).

Mynors C., *The Law of Trees, Forests and Hedgerows* (Sweet & Maxwell, 2002).

National Joint Utilities Group, *NJUG Volume 4 – Guidelines for the Planning, Installation and Maintenance of Utility Apparatus in Proximity to Trees* (NJUG, 2007).

Roads Liaison Group, *Well-Maintained Highways*, code of practice (TSO, 2005).

Roberts J., Jackson N., Smith M., *Tree Roots in the Built Environment* (TSO, 2006).

Starr C., *Woodland Management* (The Crowood Press, 2005).

Trowbridge P.J., Bassuk N.L., *Trees in the Urban Landscape* (Wiley, 2004).

Urban J., *Up By Roots* (ISA, 2008).

Wolf K.L., *Trees, Parking and Green Law: Strategies for Sustainability* (USDA, University of Washington, 2004).

Tolerance of defects is reduced as the risk to people and property increases.

chapter eight

Urban Tree Management in Private Spaces

Many of the issues faced by tree managers in public places are also faced by managers of trees in private spaces. Maybe the distinction is artificial, but there are differences when one is not taking the safety of the public into account so overtly. Often, the approach taken to avoid problems is merely to exclude the public.

COMMON FACTORS

Tree management on private property still has to wrestle with the main urban features, that is, vehicles, pedestrians, boundaries, buildings and drains. These are treated briefly below before trees on particular types of site are addressed.

Grassrings prevent soil surface problems but involve excavation and deeper soil compaction. (Photo courtesy of Greenleaf)

Vehicles

Wherever trees and vehicles are brought together, the management issues are how to keep the trees from obstructing the road or sight lines and how to prevent the vehicles from damaging the trees. Pruning of trees near to roads, whether public or private, is a routine operation to enable vehicles to pass easily and for drivers to be able to clearly see the road ahead and any signs placed at the roadside.

Keeping the trees safe from vehicles is the other side of the coin. Vehicles must be prevented from gaining access to the open ground around the tree, or this access must be provided without compromising tree roots.

Trees on roadsides are at risk of collision with vehicles and, probably, very little can be done to change this fact. Bollards and safety bars are usually designed to crumple when subjected to a collision, whereas trees are designed to resist such impact. Research in the USA shows that injuries sustained in collisions with trees are more serious than those where the vehicle hit a more yielding object (see Wolf, 2006). So, even where protection is provided between the tree and the road, trees are still likely to suffer.

An equally common problem, but one that is theoretically easier to deal with, is vehicle access to open ground containing tree roots. A single pass of a vehicle can produce compaction,

A cellular confinement system during installation.

rutting and damage to the soil. Informal parking is often seen where kerbs are absent or inadequate at a roadside. The two main options for management are to tolerate the situation, or to avoid it. Tolerance of informal parking should lead to protecting the soil from compaction and deterioration. Avoidance will lead to higher kerbs, bollards or other blocks. A tree trunk laid along the road edge is usually effective, but not always appropriate.

If informal parking is to be tolerated, the ground must be protected by a cellular confinement system that will avoid root damage and prevent soil compaction. The soil surface can be superficially protected by laying plastic grids that allow water and air percolation, but these require compaction beneath the grids and do not prevent soil compaction.

Where there are existing tree roots the best available approach at present, which is

A paviour surface over a cellular confinement system.

becoming common, is to scrape away the top vegetation (to a depth of around 50mm [2in]) and to install a cellular confinement system which spreads loads laterally. Further developments and research are needed to improve the materials and practices currently used to prevent soil compaction and to protect tree roots.

The cellular confinement system is equally useful as a temporary or permanent driveway near to trees. The main elements are indicated in the illustration below, but the principle followed is to minimize excavation, lay a geotextile to maintain a separation of layers and, on top of this, open a cellular panel. Into this panel is poured angular stone, without fine stones and soil, which will retain enough gaps to allow water and air movement. The stones fill the cells of the panel to overflowing and are then compacted into it. This forces the stones together and the plastic, interwoven cells hold the whole structure together so that loads applied on top are supported and the downward forces spread laterally over the panel, rather than being transferred through to the soil beneath. Another geotextile membrane prevents sand laid above it from dropping into the voids between the stones. Surfacing, tarmac, paviours, grass or gravel, can be added above the sand-blinding layer as a wearing course.

Pedestrians

Access for pedestrians near to trees can be provided using the same principles as above. A cellular confinement system for pedestrians still needs to be stable, but as the path widths are narrow the cellular panels need to be cut to size. Small, rigid tiles are less satisfactory, as their cells are too small for effective stone sizes and less rigid where tiles are joined. However, tiles can provide greater permeability than paving slabs and may be acceptable where installation does not cause compaction.

Path surfacing can be of gravel, bark mulch, tarmac, paviours or grass, but where disabled access is proposed the surface needs to be level, firm and stable. For single wheelchair use paths should be a minimum of 0.9m (3ft) wide, or 1.2m (4ft) to allow a pedestrian alongside. For two wheelchairs to pass, the width needs to be 1.8m (6ft). The maximum gradient for disabled pedestrians is recommended to be 1 in 15 by the Sensory Trust, which takes into account older people and those with limited upper body strength. Recommended maximum camber is 1 to 50, but this should be kept as gentle as possible for both wheelchair users and those with visual impairment. Trees benefit from cambered paths, as water is shed onto open ground where the roots can get access to it.

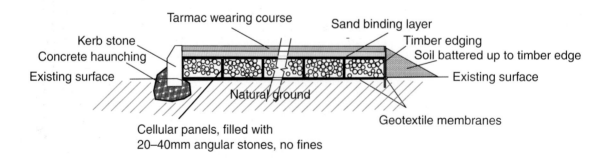

Profile of a cellular confinement system installation.

Boundaries

Trees on boundaries may cause problems via their roots, trunks and branches encroaching onto adjacent land. Hedges and trees growing as boundary screens should ideally be set back into the site by 1–2m (3–6ft), so that the trunks do not easily grow across the boundary and pruning of the outer side will not ruin the trees.

Pruning of overhanging branches is commonplace and, provided there are no planning constraints relating to trees, pruning back to the boundary can be done by the neighbour affected. This is true also for roots, but common law does not allow such work to be done if it will result in an unstable tree.

Buildings

Buildings are affected by trees through direct action, where tree growth disrupts or displaces structures, services or pipelines, or where the building is damaged by impact from branches or parts of the tree, either by natural swaying or from the structural failure of the tree.

Roots and Tree Base

The roots and base of trees disrupt buildings and structures by incremental growth where they are in close proximity. However, the forces exerted by the tree are relatively modest and only lightly loaded or unsupported structures will be affected. Free-standing walls, concrete slabs or paving stones may be affected, but tree roots do not have the strength to disrupt the foundations of heavy structures, such as a house. The greatest forces are exerted by the tree close to the base of the trunk. Here, buttress roots may form, or the gradual increase in diameter of the trunk may be very close to a structure. Also, if the trunk of the tree sways in the wind this provides additional force. Roots can also be affected by the swaying of the whole tree, increasing pressure on paving slabs, or underground pipes around which they grow.

Although tree roots have the power to cause problems, there are often a number of factors at work, including the age of the structure, the quality of materials used, water leaks and the standard of workmanship during construction. The tree may not be the main factor in the situation.

Where walls or other features have been disrupted by roots, the simple solution may be to remove the tree. But the value of the tree and the benefits that it provides should be considered first. If the tree is of high value, it may be that the wall needs to be rebuilt in a way that protects the tree and avoids the situation occurring again.

Discontinuous foundations can be used to bridge over tree roots. Lintels can be used to carry the load above the gap and, providing the roots have sufficient separation from the masonry, they can grow without causing problems. Alternatively, foundations can be reinforced to withstand the expected forces. Where only surface paving is disrupted, a possible solution may be to use a flexible surface that is more forgiving of minor distortions.

Tree roots do not break their way into underground pipes that are in good condition. There is no way that a root can bore through the pipe, as though seeking out the water within. Rather, roots grow where the conditions are favourable and so may proliferate around a broken or defective pipe. When there is a crack in the pipe, or the joint has become loose, roots can enter and form a mat of fibres that are able to block the pipe. Therefore, the way to prevent root ingress to pipes is to keep them in good condition. Where leaks are known or suspected, modern materials are available which can be threaded through the pipes and inflated to form a seamless, inner sleeve that is impervious to roots.

Inspection chambers are usually built of brick with coarse mortar, if any, and here there is a possibility of root ingress. Usually, this can be dealt with by periodic cutting of the roots, but with old chambers there may come a time when they should be rebuilt. The value of the tree involved should always be considered when seeking to find a solution.

Tree roots can also disrupt pipes by growing beneath them and then exerting an upward force as the tree sways in the wind. In this case, some root pruning may be necessary, but if the swaying extends below ground level in this way, the tree's stability should be assessed.

Trunk and Branches

Tree branches and stems are designed to sway in the wind and this can bring them into contact with structures. Pruning is usually sufficient to keep control of this situation. Where trees are protected by a TPO, local authorities routinely grant consent for cutting back branches. A 2m (6.5ft) gap between a structure and the closest branch tip is a commonly accepted specification.

New Planting

It makes sense not to plant trees too close to structures where damage may occur. But how close is too close? British Standard 5837 recommends distances that are very useful. Some other guidance has been published in newspapers in recent years recommending far greater distances. This information appears to have originated from the Kew Tree Root Survey. This study was not intended to be used as a prohibition on planting trees near structures, but simply showed how distant some trees could be while still causing indirect damage to structures on a clay subsoil.

Indirect Root Damage (Subsidence)

Trees may cause indirect damage to buildings

HOW FAR APART SHOULD YOUNG TREES AND STRUCTURES BE TO AVOID DAMAGE FROM FUTURE TREE GROWTH?

Structure type	Expected stem diameter of mature tree (at 1.5m above ground level)		
	Up to 30cm	30–60cm	>60cm
Lightly loaded structures (garages, porches etc)	No minimum distance specified	0.7m	1.5m
Heavier buildings (houses, bungalows or larger)	No minimum distance specified	0.5m	1.2m
Drains or underground services <1m deep	0.5m	1.5m	3.0m
>1m deep	No minimum distance specified	1.0m	2.0m
Masonry walls	No minimum distance specified	0.5m allows for minor movement/damage	1.0m allows for minor movement/damage
	No minimum distance specified	1.0m will generally avoid all damage	2.0m will generally avoid all damage
Concrete paths or hard standing areas	0.5m will generally avoid all damage	0.5m allows for minor movement/damage	1.5m allows for minor movement/damage
		1.0m will generally avoid all damage	2.5m will generally avoid all damage
Paths and drives with paving slabs or other flexible surface	0.7m will generally avoid all damage	0.5m allows for minor movement/damage	1.0m allows for minor movement/damage
		1.5m will generally avoid all damage	3.0m will generally avoid all damage

Based on Table 3 of BS 5837 (British Standards Institute).

when their roots draw out moisture from a clay subsoil on which foundations are resting, causing those footings to subside as the clay shrinks. There are millions of trees growing close to buildings on clay subsoils in Britain and most of them are manifestly not causing problems. But UK insurance subsidence claims often include some element of tree-related damage. Most situations require careful, professional assessment of all the relevant factors and it is inappropriate to construct an equation that adds building subsidence damage to clay subsoils, mixes in the presence of nearby trees and therefore categorically states that the trees are the cause of the problem. However, the potential for damage exists and *some* of the factors that increase the chances of damage are:

- trees growing fast and just before entering their mature stage of life
- trees growing at a distance from buildings where a large amount of water-absorbing roots are active
- foundations that are inadequate to cope with moisture removal by tree roots; that is, foundations which do not follow the guidance provided by the National House Building Council (NHBC)
- buildings composed of poor-quality materials, built to a poor standard, or inadequately maintained
- construction of areas of hardstanding that prevent soil rehydration during the winter
- the type of clay present; clays can have a range of chemical characteristics ranging from a low swelling potential to a much larger one.

All vegetation growing in a clay soil has the capability of affecting the soil volume. There are four types of ground movement associated with vegetation identified by the Building Research Establishment (*see BRE Digest* 298, 1999):

- normal seasonal movements are the result of evaporation and transpiration from the leaves of plants and ground surface
- enhanced seasonal movements are associated with an increase in transpiration when trees are introduced
- long-term subsidence results as moisture extraction by roots develops into a persistent drying pattern, with the water replaced during the winter not returning the moisture content to the original level
- long-term heave is the opposite of a persistent water deficit, with the soil progressively expanding without shrinking back to an original volume.

Where trees have been found to be implicated in subsidence damage, a decision about how to manage them is required. In some instances pruning of the crown of the tree is recommended, on the assumption that a tree with a smaller crown will need a smaller root system and draw less water from the soil. Research has found that crown reduction of trees does not effectively reduce water absorption by a tree over a sustained period unless the pruning is repeated frequently. The London Tree Officers Association (LTOA) has drawn up a strategy to identify trees growing on areas vulnerable to subsidence and regularly prune them to reduce the risks of damage from root growth. Crown thinning is an alternative pruning operation to reduction, but this has been found to have no effect on moisture content and so is not recommended.

In many cases the trees implicated in damage will need to be removed. But before this drastic action is taken, some thought must be given to the risk of heave; that is, the swelling of the soil once the tree roots are no longer drawing water from it. The important factors in a heave assessment, from an arboricultural perspective, are the age of the building and the age of the tree. If the tree was already influencing the moisture content of the clay before the building was erected, then its removal could lead to the soil swelling to a greater volume than at the time of construction. In this case, removal of the tree may cause more problems than it would solve. Where a building is older than the tree, there is much less risk of heave occurring. However, heave assessments should be carried out by qualified structural engineers.

Drains (SUDS)

Sustainable Urban Drainage Systems (SUDS) provide an alternative way of dealing with drainage on urban sites. Examples include rainwater harvesting, permeable surfacing, soakaways and ponds. This approach to water management can benefit trees by directing water to the roots, or making a store of water available close to them. Trees also are a tool to be used for soaking up such water and so becoming part of SUDS themselves.

The principle behind SUDS is to minimize the natural movement of water from a development, thereby reducing flood risk, improving water quality, providing ecological benefits and often providing attractive features that can make urban areas more desirable places to live in, while also enhancing the quality of life in the face of climate change. This approach helps to adapt the area to more extreme weather events.

Infiltration of water into the soil is a major SUDS technique for dealing with collected water, but it is not appropriate in all situations. Where infiltration is not possible or desirable, features such as permeable surfacing, swales, ponds and wetlands can still be effective. Water can be treated and led away from contaminated or compacted areas, for instance.

Conventional drainage is very disruptive to a site, involving complex planning of pipes and discharges. Drain runs need to be excavated and the material disposed of, pipes need to be laid carefully and sealed and material needs to be properly backfilled. A SUDS approach can be cheaper and have less impact on the surrounding hydrology and natural features. Land needed for SUDS can often be found by creative use of areas designated for green space, with the result that this approach does not always require additional space.

Selection and design of SUDS considers water quality and quantity equally with amenity issues. Each SUDS project is unique, being a response to the individual needs of the site. The process is multi-disciplinary, involving planning, water resources, architectural and landscape inputs.

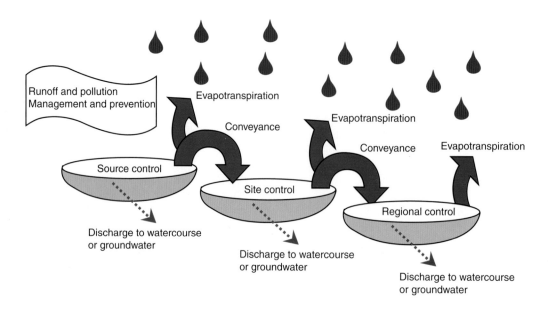

SUDS management train. (Adapted from CIRIA; www.ciria.com)

SPECIFIC FACTORS

Railways

Network Rail has responsibility for the management of over 30,000 hectares of land along the 20,000 miles of track in the UK. The Network Rail Biodiversity Action Plan informs all the decisions taken on their land and the management regimes for controlling vegetation alongside railway lines are now carefully planned and controlled. The hazards they seek to minimize are dangerous trees, or parts of trees, that could obstruct trains by falling onto the lines, or the build-up of debris that could constitute a fire hazard. For these reasons, all vegetation and trees are routinely flailed within 5m (16ft) of the tracks and kept back from the edges of the rails. Work on railways is generally carried out at night and weekends to minimize disruption to journey timetables, thus increasing the level of professionalism needed by the operatives.

The rail authority also has the right under the Regulation of Railways Act 1868 to apply to a local magistrate for an order to require the removal of dangerous trees on land next to the railway, if they constitute a danger to rail traffic. Such an order may include an instruction to the rail authority to pay compensation to the tree owner. This procedure applies to all railways, under whatever ownership. However, the normal procedure is for the rail authority to negotiate with the tree owner to get the work done. Where vegetation must be removed from sites for safety reasons, Network Rail is working with the Tree Council to identify alternative, local areas for tree replanting.

Extensions to the national rail network are likely to have impacts on the vegetation of the local areas. Environmental statements from Network Rail highlight the need to limit tree loss, identify suitable mitigating actions and to specify work when installing new infrastructure.

Airfields

Tree management on civilian airfields considers the normal factors already discussed. However, where trees grow to a significant height they

Tree growth is an ongoing issue for railway track management.

may interfere with the safe operation of the airport. Under the Civil Aviation Act 1982 (section 46), the Secretary of State can issue an order restricting the height of trees upon any land within a specified area, or requiring trees there to be cut down, or reduced in height. Any such order will need to be issued after consultation with all the affected local authorities.

Factories and Offices

Trees close to large buildings without windows can more easily be tolerated than where buildings have significant fenestration. Warehouses tend to be unsightly, functional facilities and natural growth around them can only improve their appearance. Trees, in particular, are able to provide a three-dimensional element that helps buildings to blend into the natural surroundings. However, the basic factors of access and security need to be considered. Low branches over doors and entrances need to be kept back and security fencing should not be compromised by tree growth. CCTV cameras need to maintain clear views, but there is usually some leeway in positioning cameras and, where trees are recognized as contributing to the amenities of the site, creative thinking can often find alternatives to drastic lopping, or removal of the trees.

Factories and offices usually have associated car parks and here the same issues arise as for shopping centres, colleges and hospitals of the use of trees regarding visual, environmental and psychological benefits.

Schools

Recent issues relating to trees on school grounds include health and safety of the pupils, security, provision of shading and utilization of the natural resource that trees and small areas of woodland provide.

Clearly, schools are an example of a highly emotive situation in which tolerance of tree defects is especially low. While this should not automatically result in the wholesale loss of trees from school grounds, it does reinforce the need to get trees on school grounds regularly checked and all necessary work carried out. Once this has been achieved, the focus of site managers can shift to the benefits trees provide in an informal way during playtimes and recreation breaks and the opportunities to use them as educational resources. If trees are to be seen in future as important elements in our urban life-support system, it makes sense for children to meet them as they pass through the educational system. Isolation from trees as playthings and springboards to adventure will not produce a generation ready to welcome them in the centre of cities and towns.

Recent concerns about exposure of children to sunlight have resulted in efforts to increase shading in school playgrounds. Shade protects from solar ultraviolet (UV) radiation and is particularly valuable where students are involved in outdoor activities and sports during the peak sun hours of the day, between 10am and 4pm. Shade protection can be provided by solid-roofed structures, with at least three sides open, which can also be used to protect from rain and as outdoor classrooms. But shade strategies should also include natural shade. The advantages of using trees as shade in school grounds include the reduction of ambient temperatures, low cost in comparison with permanent structures and seasonal changes to the shade, when considering deciduous trees, so that less shading and cooling is provided in winter.

Trees intended to provide shade should be to the south and west of the playground or area to be protected in order to maximize the protection during midday and afternoon hours.

Houses and Gardens

Houses and gardens provide an important resource of nature in urban areas. Britain is the

A shade tree in a school playground.

most gardenized country in Europe; there are 15 million domestic gardens in the UK. Almost one-fifth, 19.7 per cent, of London is taken up as gardens. Our relationship with our gardens is a complex one. Although many new houses are being built with just enough space at the back for a barbecue and a whirligig washing line, there are still opportunities for trees to develop. To that end, issues with trees that gardeners have faced over the decades still need to be negotiated.

Hedges

Boundaries provide almost immeasurable opportunities for disputes between neighbours and trees on boundaries can lead to challenges in managing a garden, or a neighbour.

A good hedge is relatively inexpensive to create and can be long-lasting. It is a useful weather and dust filter, can encourage wildlife and be a feature of beauty or interest, as well as providing privacy and security. A poor hedge is one that does not fulfil the role its owner intended it for, or that exceeds the space available to it, or encroaches on the land or enjoyment of a neighbour. Privacy is a major objective of homeowners, both in their houses, but also in the yard or garden, and full screening from the street or neighbouring property is often a resident's goal. Advice on designing for privacy is found in *Better Places to Live by Design* a companion guide to *Planning Policy Guidance 3 (PPG3)*.

Hedges (and tree groups or individual trees) can be established during the development of an area and should be sited to interrupt sight lines into rooms at the rear of houses and into private garden areas. The effect a hedge has on its surroundings is to a large extent dependent on its orientation and any hedge-planting should consider this. The aim of boundary hedges should be to provide a reasonable degree of privacy for neighbouring properties and to mitigate intrusive elements of a development.

There are no general restrictions on planting hedges in gardens. There are rules governing the height of boundary walls and fences, but they do not apply to hedges. However, planning conditions can be used by a local authority where they are necessary, or reasonable, precise, enforceable and relevant to the planning control of a proposed development. Such conditions may stipulate tree species that can be used in hedges and require them to be maintained at a set height. They may also be used to ensure that suitable screening and boundary treatments are carried out.

Covenants are legal restrictions on properties that are specified in title deeds. They may be used to ensure long-term retention of hedges or require that maintenance operations are carried out. They can stipulate maximum hedge heights. Developers can use covenants before new properties are sold to control how hedges are managed on a site. This could help to prevent future hedge problems.

Fast-growing species are valuable when there is a need to establish shelter in exposed locations, or where a quick screen is needed. But, if not kept under control, they can overwhelm adjacent land. Many hedge problems relate to fast-growing tree species that are difficult to maintain and out of scale with their surroundings. Choice of the right species, therefore, is an important part of achieving the desired effect and avoiding future problems.

Species that will not grow too large or too quickly are often best for hedges. Holly, yew, berberis and hornbeam are good examples. A new hedge should be given time to grow into the role for which it was planted. Possibly, a temporary screen would be useful in helping to improve privacy while the hedge is still small.

Maintenance of hedges should be considered before the hedge is planted. All hedges need regular trimming, but some need more attention than others. In general, the faster-growing species, which are large trees, are always thrusting upward and need attention more frequently than slower-growing species, which are often large shrubs rather than trees. Also, some species cope better with pruning than others. For instance, hornbeam can be severely pruned but will grow back whereas cypresses, in general, can be permanently disfigured by hard pruning.

Problem Hedges

Where hedges become troublesome it is always best to deal with the issues by discussion and agreement between the affected parties. But not everybody enjoys good relations with their neighbours. Mediation is an effective method of finding solutions to disputes. The mediator's job is to help the parties concerned to understand each other's point of view, dealing with feelings as well as facts. The mediator does not apportion blame; rather, he or she helps the parties to identify their own solution, which requires that people be willing to approach the problem in this way.

The Mediation UK website (www.mediationuk. org.uk) has further information about local community mediation services. The Community Legal Service is also a useful resource. This service maintains a list of lawyers and advice centres that offer help in hedge, as well as other, disputes. The directory can be accessed at www.communitylegaladvice.org.uk. The Citizens Advice Bureau (www.adviceguide.org. uk) gives free, confidential and independent advice and can be useful in putting people in touch with other helpful resources, or in crafting a letter to a neighbour.

Anti-Social Behaviour Act 2003

When problems relating to high hedges cannot be resolved by mutual agreement, then part 8 of the Anti-Social Behaviour Act 2003 provides power to the local planning authority to make a judgment about what action is needed. The service is not cheap as the council can charge for doing the assessment, but it does provide a way forward if all other efforts at resolution have failed. The council acts as an independent third party, in order to adjudicate about the hedge, not to mediate or negotiate with the disputing parties. This legislation, and the research carried out by the Building Research Establishment (BRE) to support it, provides a way of determining the likely effects of boundary hedges of different heights and orientations on gardens and windows.

The BRE guidance introduces the concept of

an 'action hedge height', above which a hedge is likely to block so much light that it begins significantly to affect a neighbour's enjoyment of their garden or dwelling. To be considered in this way, a hedge must consist of two or more evergreen (or semi-evergreen) trees with a joint crown and be over 2m (6.5ft) tall. The starting assumption for all the calculations is that a 2m (6.5ft) high hedge is effective as a boundary screen and also innocuous enough to have no likelihood of being a nuisance to a neighbour. The measurements and calculations lead to an action hedge height above which there will be a significant loss of light.

The extent of shade cast by a hedge depends on the height and orientation of the hedge. The greater the proportion of garden area or windows shaded, the larger will be the impact on its amenity use. More details of the method used are set out in *Hedge Height and Light Loss*, ODPM 2004.

Birds and Wildlife Encouragement

Despite the predominance of man-made structures and impact in built-up areas, urban biodiversity can be higher than in surrounding rural areas. The inclusion of 'green' urban policies over the last decade or two appears to be having an effect and, now, city gardens have

become refuges for species expelled from the countryside by industrial agriculture. The Royal Society for the Protection of Birds suggests that gardens, parks and churchyards all support a wealth of wildlife. It seems that town mouse is actually doing better in some places than his country cousin!

Species that can adapt to the presence of man are clearly going to have an easier time in cities than those which cannot. But there are clear advantages to birds in trying to make it work. In towns there are additional food sources and a potentially wider range of nesting sites. The swift and housemartin now depend on man for their nest sites, while sparrows and pigeons have become predominantly town birds. The blackbird is twice as common in towns as in the countryside. Even birds of prey are finding ways to cope with city life; using tall buildings as surrogate cliffs and feasting off the other wildlife in town.

Encouraging wildlife in a garden involves maintaining a range of textures and different site uses. Native species of plant will be recognized more easily as food by birds and animals. However, with regard to trees it is inappropriate to stick just to natives as there are only around thirty-five species of native tree in the UK. It is better to look for additional plants

Action height. Growth above this height will affect neighbours

Prune below action height to allow for growth

High hedge action height.

with wildlife potential to add to the natives.

A wildlife gardener will seek to provide suitable nesting conditions within the site, which may mean keeping some areas untidy and overgrown and working carefully when turning the compost heap, for instance. Food species to encourage should be ones that help to bridge the hungry times of year. For instance, ivy is a good source of food in late autumn when little else is fruiting. Obviously, knowing the needs of different species helps in providing for them, but if we do not know what species are in our area, a generalist approach makes sense.

Some species of wildlife are mutually exclusive, for instance frogs and fish do not coexist easily (usually the fish eat the tadpoles). Equally, a wildlife garden is unlikely to thrive at the same time as the local cat population. Cats are predators that catch and kill far more wildlife than they eat. So, excluding cats from a garden is one positive step toward encouraging wildlife.

Dos and Don'ts in the Garden

Compost piled against a tree base will start to stew, raising the temperature and moisture content around the bark. This can lead to bark damage and infection. Also, composting material uses a lot of oxygen at the early stages of the process, which can lead to a deficiency in the soil beneath, where the roots require it.

Washing lines are needed in most gardens and a strong tree is a handy post. Wrapping rope around a tree trunk or branch will damage it. If the rope is left for years it could result in a serious weakness in the tree. If trees have to be used as washing-line posts, the rope should be adjusted each year. The bark can be protected by using a narrow piece of carpet or sacking.

Homeowners may plan to add retaining walls, dig out slopes, build decking and carry out all sorts of earth-moving operations. Trees are often forgotten in such contemplation and when it comes to putting the plans into action changes of soil level around trees will lead to root damage, root loss, burying of roots or instability of the whole tree. As a rule of thumb, soil levels should not be lowered within a distance from the tree equal to four times its circumference at 1.5m (5ft) above ground level. Raising the soil level need not be a major problem provided the change is not more than 150mm (6in) and the new soil is not compacted.

Native species, like this hawthorn, have more value to wildlife than introduced species. However, introduced species can still provide wildlife benefits.

Soil should not be added close to the base of a tree, in order to avoid changing conditions around the bark.

Land preparation for lawns often includes cultivation. Use of a rotavator is very damaging for small, feeding tree roots, which are usually just under the soil surface. A spade and fork should be used to loosen the ground, the soil level above roots built up by a few centimetres and the grass grown on that.

Leisure Parks and Zoos

On any site where large numbers of the public are invited at specific times the normal factors guiding tree management need to be considered. Where entry to a site is subject to an entry fee, there is a need to keep security fences in good repair. Trees growing close to boundary fences or walls can cause damage due to natural growth of roots or branches, but nearby trees could also cause problems if they fall across the boundary and allow unauthorized entry. To counter this, pruning of trees on both sides of the boundary should be carried out so as to avoid the use of the trees as aids in crossing the boundary. Removal of all trees showing significant defects that threaten a boundary fence should also be considered.

In animal enclosures in zoos the same issue is faced, with the added health and safety issue of keeping possibly dangerous animals within a security fence. Obviously, any tree inspection or pruning works need to be carefully controlled to safeguard the arborists.

Wildlife Conservation Sites

Many wildlife conservation sites within urban areas are of small size and this physical constraint is usually a primary factor that needs to inform any management activities. Trees are large plants that can disrupt wildlife habitats by shading, fertilizing or draining the ground, or through leaf fall or collapse of old trees. Specialist ecological niches may require the complete removal of trees and if this is the

A eucalyptus trunk damaged by a tight washing line.

agreed management plan for that part of a site the benefits of trees will have to be forgone. However, in most open areas managed for nature in cities, trees provide a context for wildlife to exist in stable habitats. The three-dimensional nature of trees as individuals or groups enables separation between adjacent habitats or land ownership.

Wildlife conservation is all about habitat management. For example, protection for a bat or a bug is only effective if it includes provision for the nesting, resting, mating and hunting sites and all the various other plants and animals that are bound up in its life cycle. To miss out one element renders the whole effort meaningless, because organisms need every element to be present in order to survive successfully from one year to the next. For this reason, there is a large emphasis on management planning in wildlife sites. Much effort is put into learning about the needs of endangered species and in providing for those needs. An additional objective, especially in urban areas, is the education of people about ecology and wildlife.

Tree protection against lions in Edinburgh Zoo.

Volunteer labour is an important resource to wildlife managers and it is often deployed either to plant or remove trees. The removed trees are usually relatively young, having sprouted from seed from fecund species such as sycamore or birch.

The Wildlife and Countryside Act 1981 contains most of the regulations relating to the protection of flora and fauna and this was expanded by the introduction of the Conservation (Natural Habitats, etc.) Regulations 1994, which are the UK version of the European Habitats Directive. Enforcement of the regulations was strengthened by the introduction of the Countryside and Rights of Way Act 2000.

TREES AND DEVELOPMENT

Planning

Development of land involves change. Change can be a good thing, but it can also be a threat to trees. In the UK, the current planning system requires most changes to structures on a site to be submitted to the local planning authority for consent. The main source of current planning regulations is the Town and Country Planning Act 1990. This Act also contains the basic regulations relating to TPOs, as amended by the Town and Country Planning (Trees) Regulations, 1999 and the Town and Country Planning (Trees) (Amendment) (England) Regulations, 2008.

The term 'planning' also encompasses the process of investigating options and choosing between alternative futures for a site. So the site owner or speculator looks to the local planning authority for permission to make changes according to their vision, but this can also include management of the site to safeguard or enhance features of value. Trees may be considered as existing features, or future ones to be planted.

An arboriculturalist involved in the planning process for a site will need to use British Standard BS 5837:2005) *Trees in Relation to*

Construction – Recommendations and is likely to produce three main documents to guide the developer:

- tree survey and constraints plan
- arboricultural implications assessment or study
- arboricultural method statement.

Tree Survey

Development-site tree surveys should begin with a recent, accurate topographical plan. The tree survey should be carried out by a qualified arboriculturalist and include the following information:

- reference number to record on the survey plan
- species name (common and scientific names)
- height in metres
- stem diameter in millimetres at 1.5m (5ft) above ground level
- branch spread in metres in the four cardinal compass directions
- height from the ground to the bottom of the crown
- age class of the tree; young, semi-mature, mature, over-mature or veteran
- physiological condition of the tree, assessing the leaves, young shoots and so on as good, fair, poor or dead
- structural condition of the tree; the woody parts being assessed as good, fair, poor or dead
- preliminary management recommendations, including any need for further investigation of decay or suspected defects, or the need for a detailed assessment of the potential as a wildlife habitat
- estimated likely years remaining as a positive contribution.

Each tree should be categorized into one of four categories: R, A, B or C. This is to provide an easily recognized quality assessment. On site, the arboriculturalist begins the categorization of each tree by first considering if the tree is an R tree, meaning one that is in such poor condition that it will require removal within the next ten years regardless of any proposed changes to the site. A tree that does not fall into the R category is then assessed against the criteria for the A, B and C categories.

The tree schedule information is added to the topographical plan and valuable guidance for the architect, such as tree root-protection areas and a recommended construction exclusion zone are added, producing a tree constraints plan.

Tree Constraints

Root-protection areas Good-quality trees constrain development by requiring space for existing roots below ground and for the existing trunk and branch structure above ground. Root damage is prevalent on development sites and can compromise the ability of a tree to remain as an attractive feature once construction finishes. Trees must be given sufficient root space for stability, water and air percolation and nutrition. BS 5837 calculates the minimum root-protection area as having a radius twelve times the diameter at a height of 1.5m (5ft) above ground level. Where a circular area cannot be provided, or is inappropriate, the shape of this area (but not its size) can be modified slightly, using knowledge of where tree roots are likely to be present.

Crown spreads The crown spread of a tree needs to be respected during development and layouts that show buildings too close to the tree canopy are likely to be opposed by a knowledgeable local planning authority tree officer, because there will be immediate pressure to prune the branches back. In general, layouts that require pruning of good-quality trees to make room for a building are unlikely to be granted planning consent. Some clearance between tree and proposed building should be maintained; often 2m (6.5ft) is stipulated by planning authorities as a minimum.

Where trees are plotted that are not yet at full size, this should be considered during the

BS 5837 CASCADE CHART FOR TREE QUALITY ASSESSMENT

	R	A	B	C
Expected life (years)	Less than 10	More than 40	20–40	10–20
Quality/Value		High (substantial)	Moderate (significant)	Low
1. Arboricultural values	Seriously defective trees, expected to be lost in short term. Trees dead or in decline.	Particularly good specimens. *Essential* components of groups or of recognized arboricultural features.	Trees with impaired condition that can still be remedied.	Trees with no particular arboricultural values to lift them into a higher category.
2. Landscape values	Diseased trees that could spread infection to others.	Trees, groups or woods providing definite screening of softening in the locality, into or out of the site.	Groups or woodlands forming distinct landscape features, value being from group effect.	Trees present in groups or woods but still not forming an important landscape element.
	Low quality trees suppressed by better specimens.	Trees of particular visual importance (eg avenues or other important tree groups)	Individuals situated well into a site with little visual impact beyond.	Trees offering only temporary screening benefit.
3. Cultural values		Trees, groups or woods of significant conservation, historical, commemorative or other recognised value.	Trees with clearly identifiable conservation or other cultural benefits.	Trees of very limited conservation or other cultural benefits.
Identifying colour on plan	Dark red	Light green	Mid blue	Grey

survey and extra space given to the crown to expand so that problems due to the proximity of the tree to new buildings are minimized.

Trees and shade   Trees can, in some cases, cast shade over a site to such an extent that it can be expected to be a problem for subsequent occupiers. BS 5837 suggests, where shading is recognized as a potential problem, that it is useful to show on the plan a shaded area represented by a segment of a circle with radius from the centre of the stem equal to the mature height expected for the tree. This area should stretch from northwest to due east and indicates the total area shaded at some point during the day, but, obviously, the whole area is not shaded at any one time. Where more detailed information is needed about shading by trees on a site, the guidance in BS 8206 part 2 and BR 209 should be consulted. The tree constraints are presented

TREES AND SHADE

The British Standard BS 8206-part 2 is the primary UK reference for daylight criteria. This treats daylight as having two components; skylight and sunlight. Skylight is the diffuse light from the whole sky. Sunlight comes from direct solar beams. BR209, Site Layout planning for daylight and sunlight is the main standard to be considered at the early stages of building design but this advice is not mandatory, rather its role is to help building designers and its guidelines should be interpreted flexibly.

In cloudy climates skylight is the main source of light and so any assessment of daylight at a particular point on a building should consider the area of visible sky from that point.

Incoming sunlight brings warmth and brightness but can also cause glare and too much heat. Interiors where sunlight is expected by the occupants should receive at least 25% of the probable sunlight hours, and at least 5% in winter. Probable sunlight hours is the long-term average number of hours per year when direct sunlight would fall on unobstructed ground. In London, for example, the figure is about 1500 hours per year (BR209).

Windows to be used as the main source of light for a room need external obstructions (including trees) to be not higher than 25 degrees above the horizon, which is equivalent to a vertical sky component (VSC) of 27%. Windows that are designed to allow significant sunlight into a room should, additionally, face within 90 degrees of south.

If trees are thought to form a major obstruction above the 25 degree angle then they may affect the average daylight factor within rooms. BR209 recommends trees are ignored unless they form dense, continuous belts but there is some guidance in BRE Digest 350 which factors in the transparency of trees, which are less dense than buildings in both winter and summer.

BR209 sets out procedures for determining the important parameters in daylight and sunlight calculations. These include:

- The vertical sky component (VSC) on the outside of a window wall using a skylight indicator;
- The probable sunlight hours received by a chosen point in a building layout using a sunlight availability indicator
- The times of day and year when sunlight will be available using sunpath indicators
- Solar gain indicators are used to quantify solar radiation on a south-facing vertical wall.

These calculations are tricky and not to be attempted by the faint-hearted. Don't do them unless you have to.

to the architect in plan form to inform layout design.

Arboricultural Implications Assessment

Once a layout is available, preferably after the architect has been guided by a competent tree constraints assessment, an arboriculturalist needs to consider all the likely ways in which the proposed buildings will affect trees and vice versa. This implications assessment should consider the tree-building relationships that will be produced at the end of the project and also what issues will be faced during demolition of the existing buildings and construction of the new ones.

The assessment will need to identify where tree-protection measures are needed and what operations are likely to pose threats to retained trees. This includes any special foundations or methods of work that may be needed

if structures are proposed within tree root-protection areas, where materials will be stored and where worker facilities will be located.

Knowledge of what construction methods are to be employed is also helpful, in case special techniques or equipment are needed for lifting heavy materials onto upper floors, or for installing foundations, for instance.

The arboricultural implications assessment is the document that should accompany any planning application. It is a planning requirement that any application which could affect trees includes this information. A tree-protection plan should be included with this report, showing clearly all the trees to be retained and those to be felled. It should show where all tree protective fencing is to be erected and all ground protection where important tree roots are at risk of damage. Any areas of landscaping that can be identified

Cranes are tall and need space behind them when rotating.

should also, wherever practicable, be protected by fencing so that soil compaction is avoided.

Arboricultural Method Statement

Where there are operations planned on the site that could affect trees to be retained, these should be carried out in a way that minimizes the effects on the trees. Such operations include any unavoidable works within tree-protection areas and may involve installation of foundations, underground services, drives and paths or retaining walls and changes in soil level.

All arboricultural measures necessary to ensure the protection of trees should be

Heras panels and scaffold bracing are used here to form effective tree protection fencing.

detailed in an arboricultural method statement and kept on site, available to the project manager, so that the information is at hand when it is needed. The local planning authority may well require this information to be part of a planning application in order to be satisfied that the trees will be appropriately treated during the development.

Tree Protection

The main method of protecting trees on a development site is to erect robust, rigid, vertical fence panels, in the position shown in the tree-protection plan, before any works commence. This fencing should remain unmoved until the end of the construction phase. It should be in place before any demolition takes place, as trees can be damaged by falling debris or rampaging demolition vehicles.

Ground protection should be provided where fencing is not appropriate but work is proposed within a tree root-protection area. This has a geotextile membrane base to prevent mixing of construction debris with soil. Suitable material to provide a level surface for a pedestrian walkway is placed onto this membrane. Sand, bark mulch or small-sized gravel may be suitable. Where vehicle access is needed, this should be provided using a cellular confinement system as detailed under 'Vehicles' above.

Tree protection is also provided by following careful working methods when these must unavoidably take place close to trees. For the arboricultural guidance to be effective it must be followed by the project manager. The illustrations here show tree roots protected during the installation of garage foundations.

Underground services are a normal element in new developments. Such works can usually be planned to be installed efficiently at the start of the implementation of a project. It is important that this issue is discussed with the project manager, so that service providers do not choose their own route onto the site, ignoring tree-protection areas. Whenever services are

Exposed roots covered in hessian sacking.

Soil backfilled around roots (hessian removed).

Soil packed around roots to just below concrete supports to avoid compaction.

proposed, every effort should be made to keep them beyond any root-protection area and at all times the NJUG guidelines should be followed. There are ways of minimizing excavation, ranging from using a common trench, in which all the services are laid, to the use of trenchless technology, using thrust-boring equipment that tunnels beneath tree roots. This equipment does, however, usually require a sizeable pit at either end of the bore and these should be planned to be outside of any tree root-protection area.

Excavations for structures may expose tree roots that have to be retained, in which case the foundations will need to be designed to avoid damage to the roots. Strip foundations can be designed to bridge across roots and pads or piles can be used to minimize excavation. Separation between cement materials and roots must be maintained by use of soil or polythene sheeting and space must also be kept between the structure and the roots to allow for growth.

Landscaping

Landscaping works are usually implemented toward the end of the construction phase on a development site. With larger schemes, landscaping may be highlighted to be carried out earlier, either as a requirement by the local planning authority, or to ensure that new plants have time to settle into their new home and to add that air of maturity before the buildings are sold or occupied. It is always best to get new trees planted as quickly as practicable within a project, so long as they are not at risk of damage by the construction works, as time can never be recouped later. Although larger trees can be brought onto a site, in the long term this does not regain time lost. Larger trees typically take more time to settle into a new location and need more maintenance attention than small trees, so their growth can be affected by neglect or inappropriate works.

New trees should be planted into locations where there is adequate soil, both quantity and quality, and where drainage is effective. New developments may alter groundwater levels on a site and increasing areas of hardstanding can have unintentional effects on water availability for trees.

Where soils show signs of contamination, this should be investigated by a specialist laboratory and expert advice taken on suitable remediation measures before new trees are planted.

Ground preparation prior to shrub and tree planting or lawn laying can often involve cultivation by a rotavator. Use of this machine is inappropriate within any tree root-protection area as it is designed to churn up the superficial soil layers and chop anything there into small chunks, including tree roots. The best way to cultivate such sensitive areas is by judicious use of experienced contractors using a fork and spade.

The BRE Environmental Assessment Method

Tree surveys and assessments on development sites are now becoming measured against criteria that are designed to certify environmental standards of design and construction. BREEAM (Building Research Establishment Environmental Assessment Method) has become the most widely used environmental document for buildings in the UK and it is identified as best practice in sustainable design. The method was established in 1990 as a tool to measure the sustainability of new non-domestic buildings in the UK. In 2008 it was significantly updated, in line with UK building regulations, and is now intended to be updated periodically. The standard covers most main building types.

Ecological credits are one category of assessment within BREEAM. The process provides the opportunity to reward those schemes that contribute to enhancing biodiversity through working methods that are designed to improve living environments and meet environmental objectives.

BREEAM is required by many government departments for new construction projects and local authorities are increasingly referring to it in supplementary planning guidance. However, currently it does not comply with best arboricultural practice and it should be updated to do so.

Working with the Local Planning Authority

Planning Conditions

Where local planning authorities take seriously the protection of trees on development sites, they are likely to issue tree-protection conditions with any planning consent. These conditions may simply state that the arboricultural information provided must be followed, or they may require additional assurances that the trees have been adequately considered in relation to operations that could cause them damage. Replacement trees may be required and guidance on acceptable species, sizes and planting locations may be included.

Planning Breaches and Consequences

Trees on a development site may be protected by TPOs or by conservation area regulations. In both these cases, work to trees that is not granted consent by the local planning authority is a criminal offence and can lead to prosecution. The consequences of breaches to planning conditions are likely to be related to the seriousness of the act, or omission, but can range from withdrawal of cooperation to issuing an enforcement notice and then a stop notice by the planning authority.

REFERENCES

BSI, *BS 5837: Trees in Relation to Construction – Recommendations* (British Standards Institute, 2005).

BSI, *BS 8206-2: Lighting for Buildings – Part 2; Code of Practice for Daylighting* (British Standards Institute, 1992).

CDC, *Shade Planning for America's Schools* (National Center for Chronic Disease Prevention and Health Promotion, USA, undated).

Department of the Environment, Transport and the Regions, *Tree Preservation Order: A Guide to the Law and Good Practice* (DETR, 2000).

Department of the Environment, Transport and the Regions, Addendum to *Tree Preservation Order: A Guide to the Law and Good Practice* (DETR, 2009).

Goode D., *Desk Study: Urban Nature, in Green Infrastructure Report to the Royal Commission on Environmental Pollution* (Royal Commission on Environmental Pollution Study on Urban Environments, Well-being and Health, 2006).

LINK Consortium for Horticulture LINK Project No. 212, *Controlling Water Use of Trees to Alleviate Subsidence Risk* (BRE, 2004).

Littlefair P., *Site Layout Planning for Daylight and Sunlight – A Guide to Good Practice*, BRE report BR209 (Construction Research Communications Ltd, 1998).

Littlefair P., *Hedge Height and Light Loss* (ODPM, 2004).

London Tree Officers Association, *A Risk Limitation Strategy for Tree Root Claims* (LTOA, 2008).

Office of the Deputy Prime Minister, *High Hedges Complaints: Prevention and Cure* (ODPM, 2005).

Patch D., Holding B., *Through the Trees to Development*, APN 12 (Tree Advice Trust, 2007).

Roberts J., Jackson N., Smith M., *Tree Roots in the Built Environment* (TSO, 2006).

Sensory Trust, http://www.sensorytrust.org.uk/information/factsheets/outdoor_ip.html.

Postman's Park, London. Trees are effective at creating restful places in which to escape for a while from urban stresses.

chapter nine

Urban Tree Management and the Law

THE LEGAL FRAMEWORK AFFECTING TREES IN THE UK

This information is a very brief guide to the main legal principles and instruments affecting urban tree management, but it is always appropriate to consult the rules and regulations themselves before taking action.

In UK law, someone may be held accountable for acts or omissions that lead to a nuisance to their neighbour, to negligence that leads to damage to a neighbour, or to prosecution where a criminal act is carried out.

Nuisance is defined in law as being interference in a landowner's property by an outside party. Private nuisances are of three kinds:

- nuisance by encroachment onto a neighbour's land
- nuisance by direct physical injury to a neighbour's land
- nuisance by interference with a neighbour's quiet enjoyment of his land.

Nuisance occurs where one neighbour does something on his land that interferes with the enjoyment of neighbours on their land. If that action involves encroachment onto the adjacent land it is called trespass. Although trees grow

---

TREES CAUSING A NUISANCE

- Trees can cause nuisance when they affect a neighbour's property. You can't claim nuisance is caused by your own tree.
- There is a right to carry out work to a tree causing, or about to cause, a nuisance without the need to seek local planning authority approval. However, the authority should be informed of the proposed work.

---

from one land onto or over another, the courts have consistently classified the situation as nuisance, not trespass.

This area of the law is unusual because there is a presumption that landowners can take control of the situation and abate the nuisance using a 'self-help remedy'. In general in UK law the idea of citizens taking the law into their own hands is frowned upon.

OWNERSHIP OF TREES AND HEDGES

Unless there is evidence to suggest differently, trees belong to the owner of the land on which they grow. Tenants occupying land may not have responsibility for trees if, within the

Trees growing on boundaries tend to be pruned very hard on the boundary side.

Trees on or near Boundaries

Ownership of trees on or near to boundaries can be difficult to clarify due to lack of records, changes in boundary markers and historic decisions about the management of the trees. So, a first action in any situation where the ownership of trees is unclear must be to determine the position and ownership of the boundary. Ownership of a tree will become clear if the boundary issue is clear. If the deeds do not provide this detail, a boundary search by the Land Registry Services (www.landsearch.net) may clarify things.

In the past, boundary lines were often marked by digging a ditch and forming a bank with the excavated soil, the principle being that the soil must belong to the boundary owner and be placed on the owner's land, so the boundary is marked by the outer edge of the ditch rather than the top of the bank. This clear principle is easily complicated over time, with multiple owners, cooperative work by neighbours, internal boundaries and diverse ways of marking and dealing with boundary issues, so it is always appropriate to try to establish a boundary line by reference to authoritative documents, such as the deeds of a property.

Where boundaries are jointly owned and a tree grows close to it, the point of origin of the tree should be determined as clearly as possible, as the tree will belong to the owners of the land on which it originated. Often this is not clear and such trees will then need to be owned and managed jointly between the boundary owners.

Hedges are usually considered to be single features when trying to decide upon ownership. Individual trees may be planted on one side of a boundary or the other, but where this occurs throughout the hedge's length it suggests that the hedge is growing right along the boundary line. Obviously, if a hedge is planted slightly back from a boundary it is easier for that landowner to be recognized as the hedge owner. It is not unusual, though, for a hedge to

tenancy agreement, that responsibility has not been passed to them. But under the Occupiers' Liability Acts (1957 and 1984) and the Liability Act (Scotland) (1960), there is a duty owed by the occupier not to expose people or property to injury or damage. Within this legislation, the key concept is the degree of control that the occupier has over the premises. An occupier who has day-to-day control over a site would be held responsible for any damage or injury caused by trees growing on it.

The involvement of the local authority in tree matters is via TPOs, conservation area designation, planning conditions, or issues relating to the highway, but these constraints do not alter the basic liability and responsibility, which always lies with the tree owner.

be used as a boundary feature and ownership is then often jointly shared by the adjacent landowners.

Overhanging Branches and Roots

Trees near to boundaries often grow across them and onto neighbouring land; branches grow into the space above ground and roots encroach through the ground. There is a Legal Presumption under common law that neighbouring landowners have the right to cut back to the boundary such encroaching growths, either to abate a nuisance or to prevent obstruction of the use and enjoyment of their land, but they must keep in mind that they have a duty not to create a hazard by doing so.

So, for example, roots may be cut back to a boundary, but not to the extent of creating an unstable tree. The principle upheld by the courts seems to be that works to abate a nuisance should be kept to a minimum and so, for instance, it is inappropriate to go to court to get a tree felled because some of its branches overhang and cause a nuisance. In the court case between Leakey and the National Trust, the magistrate phrased it as 'The duty is a duty to do that which is reasonable in all the circumstances, and no more than what, if anything, is reasonable, to prevent or minimize the known risk of damage or injury to one's neighbour or to his property.' Due to the risks involved in tree pruning, it is advisable for any works to be carried out by professional, insured contractors unless they are of a very minor nature.

DEFINITION OF LEGAL PRESUMPTION

Legal Presumptions are common law rights that apply in the absence of any Act of Parliament or Contrary Agreement, such as a covenant that may supersede that right.

The parts of a tree that overhang a boundary do not at any time become the property of the landowner affected. So, branches or roots that are pruned technically still belong to the tree owner, who should not be deprived of their use. Any benefit from those severed parts should go to the tree owner and should be 'offered' back.

Boundary hedges are routinely pruned back to the boundary and this is appropriate to prevent branches extending ever further over a neighbour's land. However, a neighbour should not reach across the boundary to continue pruning, as this would become trespass. Where evergreen hedges become tall and oppressive to neighbours, the expected course of action is to discuss issues and to resolve the matter by mutual agreement. In cases where all reasonable measures to do this have failed there is the opportunity for an affected neighbour to apply to the local authority to determine if the interference with the enjoyment of his or her property is significant. The Anti-social Behaviour Act 2003 includes provision for this course of action to deal with the problem of high hedges.

HAZARDOUS TREES

A tree can cause harm because of:

- its position in relation to its surroundings
- its natural growth and the effects it has on the surroundings
- the failure of a defective part of the tree.

The first category of harm includes overhanging branches or roots which encroach onto private land or over a public highway. The second category, the natural growth of trees, refers to: the blowing down of a healthy tree in a gale; or the effects of root growth, for example in raising footpaths, cracking walls, or by modifying the ground beneath the tree, such as by allowing moss to grow or the build-up of sticky sap from aphids. The third category includes those situations in which a tree causes damage due to it being wholly or partially defective, for instance dead or decaying wood or damage to branches, trunk or roots.

LIABILITY UNDER THE OCCUPIERS' LIABILITY ACTS

The duty of care in all situations is 'to take such care as in all the circumstances of the case is reasonable'. Under the Occupiers' Liability Acts this duty is owed to visitors, neighbours and occupants of a site and even to trespassers. With regard to trees, this boils down to being aware of any risks or hazards they may pose. But at what point does 'reasonableness' come into this? At one end of the spectrum is the landowner who gets every tree inspected regularly by a professional arboriculturalist so as to become aware of the risks from the trees in detail. At the opposite end of the spectrum is the landowner who would strive to remain blissfully ignorant of any risk and who would only act when a significant hazard is pointed out. Both landowners may claim to be acting reasonably, but all landowners should know what hazards they own and take appropriate steps to reduce risks to acceptable levels.

It is recognized that no one can make their land, or their trees, completely safe, but, equally, there are some things that a prudent landowner would naturally do. These include becoming familiar with any likely sources of harm and periodically checking to see if problems have arisen, or a situation has deteriorated. Logically, any periodic checks should be more frequent where the risk of harm is clearly greatest. Where the landowner or occupier has no special expertise in tree matters, the extent of these checks is limited, but it is, nevertheless, a valid and reasonable approach to take – providing any causes for concern are taken further and someone with more specialist knowledge is consulted and their advice acted upon.

Work to or on trees can be hazardous due to the equipment and machinery used and the proximity to people and property. It is,

therefore, important to bear in mind that all operatives should be fully insured and have a duty of care to work carefully within their level of competence and to comply with all arboricultural guidelines and best practice.

TREES AND ROADS

The Highways Act 1980 requires of a highways authority that it keeps the highway clear of obstructions and safe for users. This relates to fallen stems and branches, but also to live parts of a tree that may encroach onto the road. Highways authorities, usually the district or unitary authority but sometimes the county council, generally interpret this imperative into a clearance above and to the side of the road or path surface that must be maintained. The highways authority will be responsible for maintaining the trees not in private ownership, but there is a requirement for it to be aware of any dangerous trees that could obstruct the highway by their collapse. Some authorities would actively deal with such trees themselves in order to keep the highway safe, but others issue a notice to the tree owner to deal with the issue within fourteen days. The highways authority has the power to enter land to remove a hazardous tree under the Highways Act, 1980, section 154.

TREE PRESERVATION ORDERS

It is the local planning authority's responsibility to protect important trees for the public benefit and this is achieved by the use of TPOs.

There are a number of popular misconceptions about TPOs. They are not designed to stop anyone looking after their trees and they do not mean that the local council takes on responsibility for the management or ownership of the trees.

A TPO is a legal order which gives the local authority an interest in what happens to trees on a property. TPOs protect all the live parts of a tree, including branches, leaves, trunk and roots, so when any work is proposed for the live parts of the tree the local planning authority needs to be notified.

Trees are not automatically protected. There is no law, for instance, which forbids the cutting down of oak trees, or, alternatively, that makes the felling of a non-native tree automatically acceptable. The only way a local authority can protect trees is by making a specific TPO, or, to a lesser extent, for trees to be protected by being in a conservation area or using planning conditions. Any tree which provides some measure of public benefit can be protected by a TPO. The protection comes from the TPO document, rather than from the species of tree.

All TPOs Are Not the Same

There are currently four different designations of TPO that identify protected trees in particular ways as individuals, groups, areas and woodlands. Possible changes to the planning system and regulations mean that at some point the area designation may be dropped. A TPO may include one, some or all of these categories:

- Area, or 'blanket', TPOs protect all trees that are growing on the prescribed area at the time the order is made. This is a very common type of TPO, which does not protect trees younger than the TPO, but does protect poor-quality trees that are older than it.
- A group TPO protects all the trees identified on a plan as being in a group. It may exclude some trees and it has to state how many trees are protected and their species.
- An individual tree TPO protects an individual tree, or a series of individual trees. They are plotted on a plan and their species stated.
- A woodland TPO protects an area of land as woodland. It protects all species of tree that are mentioned in the TPO document regardless of age. This means that seedlings are as protected as mature trees. Usually this type of TPO will not be used on gardens,

but there are instances where this is the case; it may be that a garden is covered by a TPO made before the area is developed.

What Work Can be Done to TPO Trees?

In England and Wales there is a unified application form for proposed work that is usually available from the local authority website. Anyone can make an application, not just the tree owner. In general, any work that maintains the safety of trees, or maintains or improves their health, is in the interests of the trees, the tree owner and the local authority. Therefore, such work is usually granted consent.

Trees do not automatically need pruning. Even a small amount of pruning causes wounds to the tree that may become infected, so there should be a clear justification for the work which indicates that the benefits outweigh the disbenefits.

Work that is proposed for a protected tree must be submitted to the council, stating what work is proposed for which tree and why. The application process takes up to eight weeks, during which time the local authority may notify neighbours of the proposed works, offering the opportunity for comment, and then a site visit is made by a council officer before the proposed work is either granted consent or refused permission and conditions are likely to be imposed.

There are situations where it is not necessary to go through this formal procedure. For example, if a neighbour's tree is causing damage it is reasonable for work to be done to abate that nuisance. The local authority should be kept informed of the work, but their consent is not needed. Trees that are imminently dangerous because of their condition do not need council

Careful pruning back of branches growing close to buildings is recognized by local planning authorities as reasonable work to protected trees. But it still needs to be presented as a tree work application.

> ### WORK WHICH IS USUALLY NOT CONTENTIOUS
>
> - Removal of dead, weak and crossing branches along with debris in the crown. This is termed 'crown cleaning'.
> - Pruning back from a building so that the tree does not damage itself when swaying in the wind and the building is not damaged by swaying branches.
> - Pruning above a driveway or public road and footpath to enable clear access
> - Ivy growing up the trunk or within a tree crown is not included in any tree preservation protection. However, it is a valuable wildlife resource and should only be removed where it is causing problems for the tree or for anyone attempting to observe the main forks and any defects in the crown and trunk.

consent either; work to make them safe can be done straight away. Again, the council should be notified of the situation, as replacement trees are automatically required for dangerous trees. Such trees may be dangerous because they are dead, seriously diseased or badly damaged. Dead parts of a tree are not protected by a TPO and can be removed without formal consent.

Works to trees on a development site which are necessary to facilitate the work are, generally, deemed to be allowed, but the work should be detailed in the approved planning documents.

The tree owner is at all times responsible for his or her tree. The TPO only confers on the local authority a power of veto when the owner proposes to carry out work that is considered to be inappropriate.

Conservation Areas

A Conservation Area is a special area designated because of architectural reasons or historical value, but the value of trees in a locality can be a factor. Regardless of the reason for the conservation area any well-established tree, with a minimum diameter at 1.5m (5ft) above ground level of 75mm (3in) (or less than 100mm [4in] where trees are removed for the benefit of adjacent trees) cannot be pruned or felled without giving the council notice.

The authority must be given six weeks' notice of intended work. However, instead of being able to refuse consent for work, the council can only respond by issuing a TPO if it objects. If this happens, the TPO regulations apply in all subsequent situations.

If the local authority does not object to the work, it may write back to say so, or may let the six-week period pass without response. If no reply has been received after six weeks, the work can proceed and must be completed within two years.

What Happens if Work is Done Without Council Consent?

It is a criminal offence for anyone to contravene the TPO or conservation area regulations, so it is important that the local planning authority is involved with any work to protected trees at an early stage and that its decision is followed.

If a tree protected by a TPO or in a conservation area is destroyed by felling or severe damage without consent, in the Magistrates' Court currently a fine of up to £20,000 can be imposed. If the case goes before the Crown Court, an unlimited fine can be imposed. The court must also consider any financial advantage gained by the removal of the tree when setting the level of fine. If the tree is damaged, but not destroyed, the Magistrates' Court can currently impose a fine of up to £2,500 per offence.

Replanting

If trees are removed following an application to the council, usually there will be a requirement to replant with another tree. The authority may ask for more than one tree to be planted for each tree felled. If a tree is felled because it is dead, dying or dangerous, there is an automatic requirement for one tree as a replacement.

What Can be Done if I Do Not Agree With the Council Decision?

If the decision of the council is contested, an appeal to the appropriate Secretary of State via the Planning Inspectorate can be made by the applicant only, provided it is lodged within twenty-eight days of receipt of the decision notice. The appeal process does not currently involve a planning fee.

An appeal is usually dealt with by written representations to the Planning Inspectorate from the appellant and the local authority. A qualified inspector then considers the submissions, visits the site, assesses the situation and prepares a report which either upholds or dismisses the appeal.

Further details of the workings of the TPO regulations are provided in Tree Preservation Orders: A Guide to the Law and Good Practice, which is downloadable from www.communities. gov.uk/publications.

STATUTORY UNDERTAKERS

Telecommunications

The Telecommunications Act, 1984, Schedule 2 provides a code of practice for work to telecommunications equipment and apparatus, including dealing with trees that are impeding the efficient operation of the system (para 19). However, it only relates to trees overhanging a street, alley, footpath or other land laid out as a *way*. Therefore, land that is away from these routes is not covered by the code. All works carried out by the telecommunications provider must be done to good arboricultural practice and in a way that minimizes damage and protects the physical environment.

The telecommunications provider can issue a notice to a tree owner, or occupier of land with trees, to require work to trees for the purpose of ensuring the efficient working of its equipment. The occupier/owner then has

twenty-eight days to comply with the notice, or to object to it. The telecommunications provider can get the work done if nothing has happened by this time. If the owner/occupier incurs expenses doing the work, they can apply to the telecommunications provider for compensation.

Electricity

Schedule 4 of the Electricity Act, 1989, provides an electricity provider with the power to carry out works to trees and shrubs to ensure the safety of the public and the electricity supply. This provision applies equally to branches above ground and roots beneath. Wherever trees cause such problems, the electricity provider is empowered to deal with them.

Where problem trees are on privately owned land, the electricity provider may issue a notice to the owner to carry out works to remove the problem. The notice should include an up-front offer to reimburse the tree owner for all reasonable work. The tree owner or occupier of the land then has twenty-one days in which to comply with the notice, or to object to the proposed works. Where there is no response from an owner/occupier, the electricity provider can get the work done without further ado.

All works should be carried out in accordance with good arboricultural practice, causing minimum damage to the trees and to the surrounding features. The Electricity Act requires electricity providers to act responsibly and to look after the beauty and amenity of the area in which they work.

Underground Pipes

The Gas Act, 1986, requires gas providers to maintain their pipes, but it does not explicitly authorize them to remove tree roots that are causing damage. It does, however, make it an offence to injure any gas fitting or pipe by neglect, which could be taken to include tree roots allowed to do the damage.

Other Statutory Undertakers

The New Roads and Streetworks Act, 1991, empowers statutory undertakers to carry out works to their apparatus where this is within highway land. In general, the Environment Act, 1995, places a duty on the Environment Agency and any authorities over which it has any power in relation to waste disposal, pollution control and land drainage to work in a way that conserves, protects and enhances the beauty or amenity of any rural or urban area.

A railway company with responsibility for the land over which the railway track runs has authority by the Regulation of Railways Act, 1868, to apply to a local magistrate for an order to remove any tree that is recognized as a hazard to the use of the track. This power only applies to land beyond the ownership of the company, as it is able to deal with its own trees without such recourse.

The Civil Aviation Act, 1982, allows an airfield authority to approach the Secretary of State for an order requiring the reduction of the height of trees on nearby land that are considered to be dangerous for the operation of the airfield. Any such order will not be issued until local planning authorities have been fully consulted.

TREE FELLING

The Forestry Commission, at the time of writing, is responsible for issuing felling licences to landowners wishing to fell trees on their land. This power is invested by the Forestry Act, 1967, for the main purpose of preserving and enhancing the amenity provided by woodlands and forests. Under Section 9 of the Act, a felling licence must be obtained to fell growing trees, but a licence is not required for pruning, trimming, lopping or topping of trees, or for the removal of dead or dying trees. The latest amendment to the 1967 Act is the Forestry (Exceptions from Restrictions of Felling) (Amendment) Regulations, 1988. The regulations controlling felling do not apply within inner London.

The felling licence regulations include numerous exemptions. Trees of diameter less than 80mm (3.2in) at 1.3m (4.25ft) above ground level, or less than 100mm (4in) where trees are removed for the benefit of adjacent trees are exempt. There is no need for a licence if the total volume of timber felled in any calendar quarter does not exceed 5m^3 (177ft^3) and the total volume sold in any calendar quarter doesn't exceed 2m^3 (70ft^3). Trees in orchards, gardens, churchyards and open spaces or those requiring felling for the prevention of danger or to abate a nuisance, do not need a licence. Exemption from the need for a felling licence does not imply that there are no other regulations which need to be complied with.

PROTECTION OF WILDLIFE

Trees are a form of wildlife, but so far in the UK only one species of tree has explicitly been protected – the Plymouth pear, *Pyrus cordata*. Old trees, large trees, or large numbers of trees together provide habitats for many species of animals and plants. The main regulations protecting such wildlife are contained in Part 1 of the Wildlife and Countryside Act, 1981 and the Conservation (Natural Habitats, etc.) Regulations, 1994, which are the UK version of the European Community Directive of Habitats and Wild Birds. Enforcement of the regulations

---

CONSERVATION AUTHORITIES

- As at 2011 the country conservation body in England is Natural England, and in Wales the Countryside Council for Wales.
- In Scotland the appropriate body is Scottish Natural Heritage under the Nature Conservation (Scotland) Act 2004 and in Northern Ireland it is the Council for Nature Conservation and the Countryside, under the Environment (Northern Ireland) Order 2002, which has powers to designate Areas of Special Scientific Interest (ASSIs).

was strengthened by the introduction of the Countryside and Rights of Way Act, 2000.

The thrust of all this legislation is to make it an offence to hurt or damage particular wildlife without a licence; works to trees may fall within this area. It is an offence to kill, injure or take any wild bird (excluding poultry and game birds), to damage or destroy its nest while under construction, or to take or destroy its eggs. There is also a list of particular birds in Schedule 1 of the Act that must not be disturbed while nesting, including the young birds, either intentionally or recklessly. Other wild animals are protected in a similar way, but without the general protection to 'all' species. Only those listed in Schedule 5 of the Act are included. Only a few of these species are likely to be relevant to normal tree management, but this list includes red squirrels and bats (all species).

In the case of bats, where proposed work could, incidentally, kill, injure or disturb them, or damage, destroy or obstruct access to bat roosts, advice from the relevant wildlife authority should be obtained. The Bat Conservation Trust provides further advice in this regard (www.bats.org.uk).

A list of wild plants is included in the Wildlife and Countryside Act in Schedule 8. It is an offence for anyone intentionally to pick, uproot or destroy any of these wild plants. It is also an offence for anyone other than an authorized person intentionally to uproot any wild plant not included in the Schedule.

When a site is designated a Site of Special Scientific Interest (SSSI) according to the Wildlife and Countryside Act, 1981, by the appropriate country conservation body, a list of operations damaging to the special features is likely to be included in the notification. These operations can, thereafter, only be carried out with written consent from the authority. The main exception would be where a detailed management plan has been agreed that includes such work. So, it is quite possible that works to trees would be included in these prohibitions. It is an offence to carry out any work that is potentially damaging without having first received written

consent from the country conservation body. The current maximum fine on conviction is £20,000.

Sites designated under international conventions or directives, such as Ramsar Convention sites, Special Protection Areas (SPAs), Special Areas of Conservation (SACs) and also land managed as national nature reserves all receive the same legal protection as SSSIs.

IMPORTANT HEDGEROWS

The Hedgerows Regulations, 1997, make it a criminal offence to remove an important hedgerow without authorization. But such a hedgerow must be either on, or adjacent to, common land, land protected as a local nature reserve or as an SSSI, land used for agriculture or forestry, or on land used for keeping or breeding horses, ponies or donkeys. Hedgerows within or bounding the curtilage of a dwelling house are specifically excluded from the regulations. The hedgerow must also be at least 20m (66ft) long or part of some other suitable length of hedgerow, or connecting with other important hedgerows at either end.

The definition of an important hedgerow is complicated, but, in summary, it must be more than thirty years old and it must meet one or more of the following criteria:

- the hedgerow marks the boundary of a historic parish or township existing before 1850
- the hedgerow contains, or is within, an archaeological feature that is on the Sites and Monuments Record, or a pre-1600 manor or estate
- the hedgerow is a part of, or associated with, a field system predating the Inclosure Acts
- the hedgerow contains species specifically protected within the Wildlife and Countryside Act, 1981, Schedules 1, 5 or 8, or various other defined species including certain Red Data Book species.

- the hedgerow is adjacent to a public right of way (not including an adopted highway) and includes at least four woody species as defined in Schedule 3 of the regulations, plus at least two associated features
- the hedgerow contains more than five woody species and has other associated, important features. The number of species required is reduced by one for some areas in the North East of England.

The full explanation of the application of these criteria and the workings of the regulations is best gleaned from *The Hedgerows Regulations: A Guide to the Law and Good Practice (The Hedgerows Guide)*.

Minor works to hedgerows that do not threaten the existence of the hedge may be exempt from the regulations. Any works that would remove part or all of an important hedgerow must be the subject of an application for approval to the local planning authority.

FINAL REMARKS

Laws and regulations are dynamic, changing goalposts and we need to remain alert if we are to avoid scoring an own goal in the future. It is always appropriate to check with knowledgeable persons and with primary sources of information before carrying out work to a tree that may be prohibited.

REFERENCES

Department of the Environment, Ministry of Agriculture, Fisheries and Food, and the Welsh Office, *The Hedgerows Regulations: A Guide to the Law and Good Practice (The Hedgerows Guide)* (DOE/MAFF, 1998).

DETR, *Tree Preservation Orders: A Guide to the Law and Good Practice* (DETR, 2000).

Mynors C., *The Law of Trees, Forests and Hedgerows* (Sweet and Maxwell, 2002).

Naturenet, http://www.naturenet.net/trees/hedgerow/.

LEFT: Dawn redwood at the rear of the Royal Courts of Justice, London.

OVERLEAF: Bloomsbury Square, London.

chapter ten

Urban Tree Management and the Future

Cities will continue to grow, spread, draw people into them and affect the climate of their surroundings. Despite recent trends in electronic technology allowing remote working, most people still need to congregate together in order to get work done and to interact with each other. City life is here to stay and more than 50 per cent of the world's population now lives in cities, so we need seriously to apply ourselves to making city life bearable. Trees can be great allies of urban life; they improve health and well-being for people and the environment, they protect from extreme temperatures, reduce pollution and increase land values.

TREE MANAGEMENT TRENDS AND TOOLS

Arboriculture in the UK seems to be going in two directions at once: it is slowly gaining ground as a scientific discipline; and awareness of the crucial role of urban trees is also growing. Urban initiatives are drawing on the considerable experience of arboricultural officers and practitioners, while technical and regulatory tools are also being developed to help with assessment and monitoring of existing trees.

Over the next decade, tools to help to determine the stress of individual urban trees should become commonplace: methods of decay diagnosis should improve, including thermal imaging becoming more sophisticated; tree radar will become an effective way to map the underground root systems of trees, particularly beneath hard surfaces; and continuing development of geographic information systems (GIS) should make inventories straightforward, so that more urban areas will have quantified assessments of their canopy cover and condition of the tree stock. These are all steps forward, but arboriculture could also slip back if the industry does not become more united and focused on its role to provide professional support in urban areas. Economic constraints will challenge local authority tree teams just to remain effective, let alone get their tree resource properly mapped.

The danger is always that the small band of tree professionals in the UK will become subsumed into some larger 'green' umbrella institution. To avoid this, we need to punch above our weight and to demonstrate professionalism and innovation to fellow professionals, earning our right to influence policy and practice throughout the rest of the century.

THE EFFECTS OF CLIMATE CHANGE

The consensus of scientific opinion is that climate change is happening. Whatever the cause of the changes currently being observed, their effects are already being felt in urban areas and are expected to become more pronounced during this century. The expected climate change effects in the UK are:

- hotter summers, but less warming effect in winter
- heavier winter rainfall
- flooding to be commonplace
- decrease in summer rainfall
- droughts to be more common.

It is worth noting that the majority of native tree species are expected to cope reasonably well with climate change.

Urban change is as great a challenge as climate change. Urban areas experience both together and they can be difficult to distinguish. Cities create their own climate and by replacing the natural surface characteristics of the land with artificial structures built of artificial materials the result is less green cover and a pronounced effect on the physical environment. Green space in urban areas affects the whole urban environment and also its interactions with the surrounding landscape. Urban design can modify the urban climate by using green space and trees as physical tools and architectural features.

Adaptation strategies in relation to trees include matching species to the expected climate of sites and changes to management practices. Immediate management measures for woodlands that can be followed include planting mixed stands of trees rather than monocultures and maintaining tree cover over an area, so that felling only ever involves small groups of trees. Woods on floodplains can help to alleviate flooding by reducing peak flows during flood events. Woods beside rivers reduce bankside erosion, improve water quality and help to maintain fish stocks by reducing water temperature.

Trees are a key element of any urban climate change adaptation strategy. They can be integrated into the urban fabric so as to provide essential benefits such as shading and cooling. Without these environmental services, cities of the future are likely to be very inhospitable places.

Trees are an essential component of strategies to limit the urban heat-island effect. Compared to other landscape features, trees provide more leaf area and cooling effect, while at the same time taking up very little area at street level; the challenge is more how to provide adequate root space. Trees leave the street, square or open space available for other uses, many of which positively benefit from the tree cover.

Increasing the tree canopy cover in cities is a cost-effective way of helping them to remain bearable for the urban population. However, the greatest benefits are associated with large individual trees or large numbers of trees and construction of all types and at all scales needs to make room, in both design and implementation, for the planting and care of these trees.

TRENDS IN URBAN LIFE

High-density living requires services and infrastructure that do not easily allow for trees. Climate change adds to these pressures. Trees are under pressure at three levels: underground; at street level; and at canopy level.

When well designed and maintained, parks and open spaces can lift adjacent property values. Good-quality tree planting and landscaping are seen as vital elements in the benefits recognized by residents and enterprises. The *Park Life Report*, by Greenspace, found that 97 per cent of people view parks as benefiting the local community.

The recognition of the role of trees in maintaining naturalness and husbanding

Large, three-dimensional features like trees help to create oases of calm.

ecology in built-up areas is gaining ground and TDAG is campaigning to ensure it will be a material consideration in the development process in the future.

POLICY FRAMEWORK

Tree lifespans can be far greater than the design life of new developments, which are down to around thirty years in non-residential districts. Our towns include trees engulfed by Victorian and Edwardian expansion, yet they have survived multiple generations of buildings and surrounding structures. We now need to build such resilience into a new generation of trees. To achieve this requires an application of all the tools, materials and design concepts currently known to us and a determination to improve on them. It can be done.

Without such an attitude, we will bequeath to our children and grandchildren an impoverished landscape that is unable to soften the harsh realities of city life and is ineffective as an urban life-support system.

Trees need to be seen as essential components of urban infrastructure and, on new developments, should be given equal priority with other services, such as drainage and underground services, security and access. A collaborative approach is needed to allow large-stature trees to be factored into the design, implementation and future functions of sites. Only when trees and green infrastructure

Trees surrounded by roads, buildings and people need careful management but they are vital elements in helping to make urban areas liveable.

are at the centre of city planning and growth will the benefits that trees provide be fully understood and appreciated.

Without such a shift in the perception of allied professionals (a paradigm shift), trees will not be able to take a full role as the effective urban life-support system that they truly are.

REFERENCES

Britt C., Johnston M., *Trees in Towns II: A New Survey of Urban Trees in England and their Condition and Management* (Department for Communities and Local Government, 2008).

CABE Space, *Does Money Grow on Trees?* (CABE, 2005).

Greenspace, *The Parklife Report* (Greenspace, 2007).

NTSG, *Bringing Common Sense to Tree Management* (NTSG, 2010).

TDAG, *No Trees, No Future* (TDAG, 2008).

The London Assembly, Environment Committee, *Chainsaw Massacre: A Review of London's Street Trees* (The London Assembly, 2007).

Townsend M., *Trees, Woods and Climate Adaptation: The Opportunities and Need for Action*, 43rd Arboricultural Association National Conference (Arboricultural Association, 2009).

Edinburgh tree views.

Useful Websites

www.adviceguide.org.uk Citizens Advice Bureau website.

www.flac.uk.com/wp-content/uploads/2009/12/TEMPO-GN.pdf TEMPO guidance notes.

www.gohelios.co.uk for National Plant Specification and for UK Plant Selector database

www.green-space.org.uk Greenspace is a registered charity which works to improve parks and green spaces by raising awareness, involving communities and creating skilled professionals.

www.Homeoffice.gov.uk/documents/besafebesecure.pdf Home Office website advice on personal safety.

www.landsearch.net Land Registry Services

www.lantra.co.uk Lantra, sector skills council for arboriculture.

www.mediationuk.org.uk The Mediation UK website has information about local community mediation services.

www.metoffice.gov.uk/climate/uk/about/archives.html for information about weather trends from weather stations throughout the UK.

www.naturenet.net/trees/hedgerow Naturenet, a popular UK independent countryside and conservation website.

www.NTSG.org.uk The National Tree Safety Group provides tree risk-management guidance.

www.right-trees.org.uk Website to aid search for appropriate species. Registration needed before access is provided.

www.thebigtreeplant.direct.gov.uk The Big Tree Plant is a campaign to encourage people and communities to plant more trees in England's towns, cities and neighbourhoods. It is a partnership bringing together national tree-planting organizations and local groups with Defra and the Forestry Commission.

www.treecall.co.uk The author's website.

www.treecouncil.org.uk UK charity promoting the importance of trees in a changing environment.

www.treeregister.org The Tree Register is a registered charity collating and updating a database of notable trees throughout Britain and Ireland.

www.trees.org.uk The Arboricultural Association is the largest amenity tree care body in the UK.

Index

RELATED TITLES
FROM CROWOOD

British Oaks

MICHAEL TYLER

ISBN 978 1 84797 041 1
240pp
200 illustrations

Hedges and Hedgelaying

MURRAY MACLEAN

ISBN 978 1 86126 868 6
200pp
200 illustrations

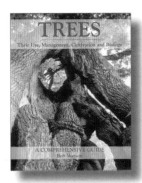

Trees

BOB WATSON

ISBN 978 1 86126 885 3
384pp
300 illustrations

Woodland Management

CHRIS STARR

ISBN 978 1 86126 789 4
192pp
120 illustrations

In case of difficulty in ordering, contact the Sales Office:

> The Crowood Press Ltd
> Ramsbury
> Wiltshire
> SN8 2HR
> UK

Tel: 44 (0) 1672 520320
enquiries@crowood.com
www.crowood.com